Optical Thin Film Design

Optical Thin Film Design

Andrew Sarangan

CRC Press

Taylor & Francis Group

Boca Raton London New York

CRC Press is an imprint of the
Taylor & Francis Group, an **informa** business

First edition published 2021
by CRC Press
6000 Broken Sound Parkway NW, Suite 300, Boca Raton, FL 33487-2742

and by CRC Press
2 Park Square, Milton Park, Abingdon, Oxon, OX14 4RN

© 2021 Taylor & Francis Group, LLC

CRC Press is an imprint of Taylor & Francis Group, LLC

ISBN: 978-1-138-39044-7 (hbk)
ISBN: 978-0-367-51271-2 (pbk)
ISBN: 978-0-429-42335-2 (ebk)

Typeset in Times
by codeMantra

Contents

Preface

This book is the result of several courses in thin films, photonics, and nanofabrication that I have developed and taught at the University of Dayton, Ohio. At first glance, the design of optical thin films might appear to be an old topic. Indeed, the basic principles of optical coatings have not changed. But with the increased penetration of photonics into virtually all aspects of consumer electronics, the application space has expanded far beyond the traditional antireflection coatings and filters. Whether touch screen devices, smart mirrors, smart windows, or laser beam scanners, all of them use a large number of optical thin films, performing complex functions that remain transparent to the end user. Hence, there has been a renewed interest in this field to serve the needs of these emerging applications.

Since there are already a number of well-known books on the subject, it is worth highlighting what is different about this book. First, I have developed this book as a pedagogical tool rather than as a reference tool. In other words, this is not a review of the latest developments in optical thin film science, nor a discussion of novel techniques used only in research laboratories. My goal is to guide the reader through the fundamental principles of thin film design using visualizations and computational methods. Second, with the exception of the first chapter where I derive Maxwell's wave equations and the boundary conditions, I have introduced every new thin film concept through numerical examples rather than through generalized derivations. This has been done to give the reader the practical context and relevance of each method. Third, this book approaches the topic of complex thin films (i.e., those that have complex refractive indices) in a more systematic and self-consistent manner than other sources. This includes metal-insulator-metal (MIM) structures and resonant-cavity enhancement (RCE) structures. Fourth, I have a whole chapter devoted to phase change materials (PCM), a topic that has generated quite a lot of interest in recent years and has many interesting applications. Fifth, every result shown in the book (there are over 200 figures of calculated results) has been computed from scratch. None of them are reproductions from other sources. Finally, the book also includes some computer codes in Python programming language which allow the reader to recreate and modify the example calculations shown in the book. These are given in the last chapter of the book.

The book is aimed primarily at graduate students who are learning the design principles of optical thin films as part of their research. Even though everything is derived from first principles, it is assumed that the reader has some prior background in electromagnetics. Since this is the first edition of this book, I don't expect it will be perfect. But I hope the main concepts come across clearly and concisely and that the reader finds this as useful and enjoyable a book as much as I enjoyed writing it.

Andrew Sarangan, PhD, PE
University of Dayton
January 2020

Author

Dr. Andrew Sarangan is a Full Professor in the Electro-Optics & Photonics Department at the University of Dayton, Ohio. He received his BASc and PhD degrees from the University of Waterloo in Canada in 1991 and 1997, respectively. His current research areas are in photodetector technologies, optical thin films, nanofabrication, nanostructured thin films, and computational electromagnetics. At Dayton, he created a state-of-the-art and comprehensive nanofabrication laboratory for thin films, lithography, and semiconductor processing, as a single-PI (principal investigator) effort from externally funded research. His research has been sponsored by the National Science Foundation, various agencies of the Department of Defense including the Air Force Research Laboratory.

1 Fundamental Concepts

1.1 OPTICAL THIN FILMS

Thin film coatings are ubiquitous on all optical components such as lenses, mirrors, cameras, and windows. The design can be as simple as a single dielectric antireflection film, or very complex with several hundred layers of films for producing elaborate optical filtering effects. A typical design will consist of a number of layers deposited sequentially on a substrate as shown in Figure 1.1, on one side of the substrate or on both sides depending on the specific application. The number of layers will depend greatly on the substrate material, desired spectral features, and the availability of coating materials. While transparent dielectrics are the most widely used thin film materials, metals and semiconductors can also be used where additional features such as electrical conductivity and broadband reflectivity are desired.

The design process involves exploiting the fundamental concepts described in the remainder of this book to come up with a layer structure that will approximately yield the desired spectral characteristics. The design method will be driven by the type of spectral characteristics desired (i.e., whether it is an antireflection, bandpass, long-pass, etc.). The desired performance at specific wavelengths, or at a range of wavelengths, will also have to be defined. This will require knowledge of the thin film materials that can be used for implementing the design. Once a structure that closely resembles the desired performance is determined, it is then used as a starting point to numerically refine and improve the performance across the entire spectral range of interest.

With the wide availability of numerically driven thin film design software, the temptation is to allow the computer to synthesize the entire film structure. While this may sometimes yield usable designs, a fundamental understanding of the basic building blocks and the design process will result in the simplest designs for the required performance, as well as create an in-depth understanding of the general principles of

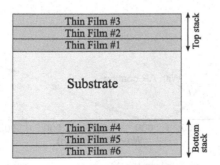

Figure 1.1   A typical optical thin film structure deposited on both sides of a substrate.

thin film design. Numerical refinement will still be necessary to optimize the design across all wavelengths, but providing the computer with a good starting point makes it possible to achieve the objectives of the design with the fewest number of thin film layers.

1.1.1 ANTIREFLECTION

Antireflection coatings are widely used on refractive optics and windows. For example, eye glasses utilize two or more layers of thin dielectric films to cut down the reflection from ~7% to <0.1%. A typical optical system such as a camera or telescope consists of a number of lenses stacked together. With 7% reflection from each lens, it is easy to see that it doesn't take much before the reflection starts to cause significant problems. The spectral bandwidth of the antireflection also has to be considered. For visible light applications, the bandwidth has to cover the wavelength range from 400 to 700 nm. The angular sensitivity is also an important consideration. The range of incident angles greatly depends on the specific application.

Antireflection coatings are also widely utilized in image sensors and photovoltaic components. The reflection from a silicon surface without any coating is about 35% in the visible range. But with just one dielectric layer, this reflection could be significantly reduced to below 0.1% as shown in Figure 1.2. The spectral range of visible image sensors is 400–700 nm, with the angles of incidence determined largely by the f/# (numerical aperture) of the optics. The spectral range of photovoltaic devices is much greater, spanning the entire detection band of silicon, from 400 to 1,100 nm. Furthermore, the angles of incidence will also be larger, with as much as ±80° to account for varying solar illumination angles.

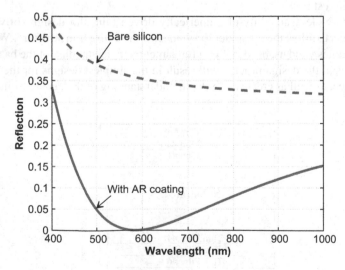

Figure 1.2 Reflection spectra at normal incidence of a silicon substrate with and without an antireflection (silicon nitride) film.

It is also possible to create antireflection on metals and other highly reflective surfaces. These are of interest in electronic components where reflected light from conductive metal interconnects can cause problems. Creating antireflection on metals can, in some cases, allow these metals to be used as partially transparent electrically conductive thin films.

1.1.2 HIGH REFLECTION

In addition to antireflection, dielectric films can also be used to produce high reflectors. Reflectivity values as high as 99.99% can be obtained by periodically alternating high and low refractive index films in a stack. Compared to metal surfaces, dielectric reflectors can provide better control of the reflection bandwidth. They can also produce lower absorption, which is important in high power laser applications. Additionally, dielectric reflectors can be used in resonant cavities and are especially attractive in semiconductor laser applications. For example, vertical cavity surface emitting lasers (VCSELs) utilize these reflectors to produce the required cavity resonance for lasing.

1.1.3 OPTICAL FILTERS

Besides antireflection and high-reflection filters, there are many other types of optical filters, such as long-pass, short-pass, bandpass, and line-pass filters. Long-pass filters transmit wavelengths longer than a certain value (known as the cut-on wavelength) and reflect shorter wavelengths. Short-pass filters do the opposite – they transmit everything shorter than the cut-off wavelength and reflect longer wavelengths. Bandpass filters transmit a band of wavelength and reflect everything outside that range. Line-pass filters are similar to bandpass filters, except their transmission band is much narrower, often coinciding with common laser wavelengths. An example is shown in Figure 1.3. All of these filters can be synthesized by combining the antireflection concepts with high-reflection concepts such that the reflection and transmission spectra exhibit both of these characteristics at different wavelength ranges. For example, a long-pass filter utilizes a high reflector designed for short wavelengths, and an antireflector designed for longer wavelengths, such that the two bands are adjacent to each other without interfering with each other. A bandpass filter utilizes an antireflection region with two high reflecting regions on either side of the antireflection region.

1.1.4 OPTICAL FILTERS WITH METAL FILMS

When metal films are made ultrathin, they can become transparent. The transparency depends entirely on the characteristics of the metal. For example, silver and gold can be very transparent, while a metal like chromium is very opaque. Ultrathin transparent metal films can be used in applications where electrical conductivity, optical transparency, and reflectivity are simultaneously desired, such as in displays and smart mirrors. Compared to the current generation of displays that use transparent

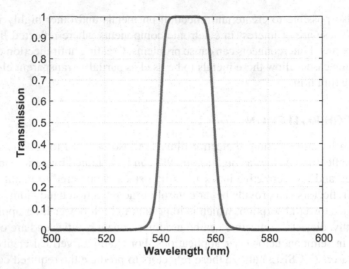

Figure 1.3 Transmission spectra at normal incidence of a narrow bandpass filter designed for 632 nm wavelength.

conductors like indium tin oxide (ITO), metal-dielectric structures have the potential for significantly higher conductivity because they contain real metals instead of partially conducting oxides. Furthermore, metal-dielectric structures can also be used as inexpensive hot mirrors (mirrors that reflect infrared and transmit visible light). Also known as low-emissive coatings, these are widely utilized in architectural windows for improving the energy efficiency of buildings. They are also used in applications that require reflection and transmission with the minimum number of layers.

Metals have complex refractive indices and therefore exhibit significantly different optical properties than dielectrics. In some metals, the imaginary part of the refractive index can be significantly larger than the real part. As a result, they can perform certain functions that are difficult to achieve with dielectrics alone. However, this also makes them absorptive, making them unsuitable in applications where high incident energies are involved, such as in high-power lasers.

1.2 ELECTROMAGNETIC WAVE EQUATION

The propagation of electromagnetic waves is a vast subject area. Fortunately, for thin film applications considered in this book, we do not need dive into the complexities of general wave propagation in three-dimensional space. Since optical thin film structures are typically produced as stratified layers on a flat substrate, we will be limiting our discussion to *plane waves* only. As shown in Figure 1.4, these are waves that have parallel fronts and appear to arrive from a source at an infinite distance away. This allows us to ignore the effects of diffraction and treat it as a one-dimensional wave propagation, greatly simplifying the analysis.

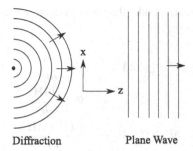

Diffraction   Plane Wave

Figure 1.4   Illustration of diffraction from a point source and a plane wave.

We will start by deriving the plane wave equation from Maxwell's electromagnetic equations. The derivation is useful to gain some insights into the origins of optical wave propagation. Then we will examine the properties of plane waves at the interface between two dielectric materials.

The four basic electromagnetic equations are

$$\nabla \times \mathbf{E} = -\frac{\partial \mathbf{B}}{\partial t} \tag{1.1}$$

$$\nabla \times \mathbf{H} = \mathbf{J} + \frac{\partial \mathbf{D}}{\partial t} \tag{1.2}$$

$$\nabla \cdot \mathbf{B} = 0 \tag{1.3}$$

$$\nabla \cdot \mathbf{D} = \rho, \tag{1.4}$$

and the two additional relations that bring the material properties into consideration are

$$\mathbf{B} = \mu \mathbf{H} \tag{1.5}$$

$$\mathbf{D} = \varepsilon \mathbf{E}. \tag{1.6}$$

E and **H** are electric and magnetic field amplitudes, while **D** and **B** are electric and magnetic flux densities. The relation between field amplitudes and flux densities are determined by the two material parameters μ (magnetic permeability) and ε (electric permittivity). Their values in vacuum are

$$\mu_0 = 4\pi \times 10^{-7} \text{ H/m} \tag{1.7}$$

$$\varepsilon_0 = 8.854 \times 10^{-12} \text{ F/m}. \tag{1.8}$$

If μ and ε are independent of field amplitudes, then the medium is said to be linear. If they are independent of the field directions, the medium is isotropic. The materials we will deal with in this book are mostly linear and isotropic. The value for μ in most thin film materials can be taken to be the same as μ_o, except in some magnetic materials such as iron, cobalt, and nickel. ε, however, varies greatly between materials and will be the main focus in our study.

To derive the wave equation, take the curl of equation (1.1) and use equation (1.5) to replace **B** with **H**. This results in

$$\nabla \times (\nabla \times \mathbf{E}) = -\nabla \times \frac{\partial \mathbf{B}}{\partial t} \tag{1.9}$$

$$= -\nabla \times \frac{\partial \mu \mathbf{H}}{\partial t} \tag{1.10}$$

$$\nabla \times (\nabla \times \mathbf{E}) = -\frac{\partial (\nabla \times \mu \mathbf{H})}{\partial t}. \tag{1.11}$$

Since $\mu = \mu_0$, its value will be independent of time and space. This results in

$$\nabla \times (\nabla \times \mathbf{E}) = -\mu_0 \frac{\partial (\nabla \times \mathbf{H})}{\partial t}. \tag{1.12}$$

Using equation (1.2) to replace **H** with **D**, and equation (1.6) to replace **D** with **E**, results in

$$\nabla \times (\nabla \times \mathbf{E}) = -\mu_0 \frac{\partial}{\partial t} \left[\frac{\partial \mathbf{D}}{\partial t} + \mathbf{J} \right] \tag{1.13}$$

$$= -\mu_0 \left[\frac{\partial^2 \varepsilon \mathbf{E}}{\partial t^2} + \frac{\partial \mathbf{J}}{\partial t} \right] \tag{1.14}$$

$$= -\mu_0 \left[\frac{\partial}{\partial t} \left(\frac{\partial \varepsilon \mathbf{E}}{\partial t} \right) + \frac{\partial \mathbf{J}}{\partial t} \right] \tag{1.15}$$

$$= -\mu_0 \left[\frac{\partial}{\partial t} \left(\mathbf{E} \frac{\partial \varepsilon}{\partial t} + \varepsilon \frac{\partial \mathbf{E}}{\partial t} \right) + \frac{\partial \mathbf{J}}{\partial t} \right] \tag{1.16}$$

$$= -\mu_0 \left[\mathbf{E} \frac{\partial^2 \varepsilon}{\partial t^2} + 2 \frac{\partial \varepsilon}{\partial t} \frac{\partial \mathbf{E}}{\partial t} + \varepsilon \frac{\partial^2 \mathbf{E}}{\partial t^2} + \frac{\partial \mathbf{J}}{\partial t} \right]. \tag{1.17}$$

The time-dependent dielectric constant represents the temporal response of the material (which can also be characterized as spectral dispersion). Ignoring dispersion for now, we can get

$$\nabla \times (\nabla \times \mathbf{E}) = -\mu_0 \left[\varepsilon \frac{\partial^2 \mathbf{E}}{\partial t^2} + \frac{\partial \mathbf{J}}{\partial t} \right]. \tag{1.18}$$

Using the mathematical identity

$$\nabla \times (\nabla \times \mathbf{E}) = \nabla (\nabla \cdot \mathbf{E}) - \nabla^2 \mathbf{E}, \tag{1.19}$$

equation (1.18) can be written as

$$\nabla^2 \mathbf{E} - \nabla (\nabla \cdot \mathbf{E}) = \mu_0 \varepsilon \frac{\partial^2 \mathbf{E}}{\partial t^2} + \mu \frac{\partial \mathbf{J}}{\partial t}. \tag{1.20}$$

We will assume that the medium has no free charges. Therefore, equation (1.4) becomes

$$\nabla \cdot \mathbf{D} = 0, \tag{1.21}$$

$$\nabla \cdot \varepsilon \mathbf{E} = 0, \tag{1.22}$$

$$\nabla \varepsilon \cdot \mathbf{E} + \varepsilon \nabla \cdot \mathbf{E} = 0, \tag{1.23}$$

$$\nabla \cdot \mathbf{E} = -\frac{\mathbf{E} \cdot \nabla \varepsilon}{\varepsilon}. \tag{1.24}$$

Substituting equation (1.24) into (1.20) produces

$$\nabla^2 \mathbf{E} - \mu \varepsilon \frac{\partial^2 \mathbf{E}}{\partial t^2} - \mu_0 \frac{\partial \mathbf{J}}{\partial t} = -\nabla \left(\frac{\mathbf{E} \cdot \nabla \varepsilon}{\varepsilon} \right), \tag{1.25}$$

which is the most general form of the electric field wave equation.

It may be curious that we did not utilize equation (1.3) in this derivation. This does not mean we have ignored Gauss's law. We can show that equation (1.3) is implicitly accounted for in equation (1.1). Take the divergence of equation (1.1) on both sides, such as

$$\nabla \cdot (\nabla \times \mathbf{E}) = \nabla \cdot \left(-\frac{\partial \mathbf{B}}{\partial t} \right) \tag{1.26}$$

$$= -\frac{\partial (\nabla \cdot \mathbf{B})}{\partial t}. \tag{1.27}$$

The left-hand side of the equation is a divergence of a curl, which is always zero. Therefore, assuming we are not dealing with static fields, we can conclude that the divergence $\nabla \cdot \mathbf{B}$ must also be zero. As a result, equation (1.3) is implicitly satisfied by equation (1.1).

We will assume that the source of the electromagnetic wave is outside the computation region. Therefore, equation (1.25) becomes strictly a propagation problem. Moreover, we will assume that the fields have a harmonic time dependence of the form $e^{j\omega t}$, such that

$$\mathbf{E} = \hat{\mathbf{E}} e^{j\omega t}, \tag{1.28}$$

where $\hat{\mathbf{E}}$ is the amplitude of the electric field vector. Since it is a linear system, the current density $\mathbf{J}$ will also have the same harmonic time dependence of the form

$$\mathbf{J} e^{j\omega t} = \sigma \hat{\mathbf{E}} e^{j\omega t}, \tag{1.29}$$

where σ is the conductivity of the material. This assumption of time dependence does not limit the generality of our approach, because a linear sum of these basic harmonics can be used to express any function of time.

In the following equations, we will use $\mathbf{E}$ instead of $\hat{\mathbf{E}}$ to unclutter the notation. With this assumed time dependence, equation (1.25) becomes

$$\nabla^2 \mathbf{E} + \mu_0 \varepsilon \omega^2 \mathbf{E} - j\omega \mu_0 \sigma \mathbf{E} = -\nabla \left(\frac{\mathbf{E} \cdot \nabla \varepsilon}{\varepsilon} \right) \tag{1.30}$$

$$\nabla^2 \mathbf{E} + \mu_0 \omega^2 \left(\varepsilon - j\frac{\sigma}{\omega} \right) \mathbf{E} = -\nabla \left(\frac{\mathbf{E} \cdot \nabla \varepsilon}{\varepsilon} \right). \tag{1.31}$$

We can consider the term $\left(\varepsilon - j\frac{\sigma}{\omega} \right)$ as the complex permittivity. Additionally, we can also define a *complex relative permittivity* ε_c as

$$\varepsilon_c = \frac{1}{\varepsilon_0} \left(\varepsilon - j\frac{\sigma}{\omega} \right) \tag{1.32}$$

$$= \frac{\varepsilon}{\varepsilon_0} - j\frac{\sigma}{\omega \varepsilon_0}. \tag{1.33}$$

This allows us to write equation (1.31) as

$$\nabla^2 \mathbf{E} + \mu_0 \varepsilon_0 \varepsilon_c \omega^2 \mathbf{E} = -\nabla \left(\frac{\mathbf{E} \cdot \nabla \varepsilon}{\varepsilon} \right). \tag{1.34}$$

Therefore, the refractive index of the material becomes

$$n = \sqrt{\varepsilon_c} = \sqrt{\frac{\varepsilon}{\varepsilon_0} - j\frac{\sigma}{\omega \varepsilon_0}} \tag{1.35}$$

$$= m - j\kappa, \tag{1.36}$$

where m is the real part and κ is the imaginary part. We can relate the real and imaginary parts of the complex permittivity and the refractive index as

$$\Re\{\varepsilon_c\} = m^2 - \kappa^2 \tag{1.37}$$

$$\Im\{\varepsilon_c\} = 2m\kappa. \tag{1.38}$$

The second coefficient of $\mathbf{E}$ in equation (1.34) can be lumped into a single parameter k^2 such as

$$k^2 = \mu_0 \varepsilon_0 \varepsilon_c \omega^2. \tag{1.39}$$

k has the units of radians/length or angular wave number. Using the free space value for ε_0, we can define a free space angular wave number k_0 as

$$k_0^2 = \mu_0 \varepsilon_0 \omega^2, \tag{1.40}$$

where μ_0 and ε_0 are the free space magnetic permeability and dielectric permittivity, respectively. As a result, equation (1.34) can be written as

$$\boxed{\nabla^2 \mathbf{E} + k_0^2 n^2 \mathbf{E} = -\nabla \left(\frac{\mathbf{E} \cdot \nabla \varepsilon_c}{\varepsilon_c} \right),} \tag{1.41}$$

which will become the basic wave equation for all of our subsequent analyses. Since k_0 is the number of radians per unit length in free space, we can define the free space distance equivalent to 2π radians as the free space wavelength, such as

$$\lambda = \frac{2\pi}{k_0}. \tag{1.42}$$

In a perfect insulator far from any resonances, ε_c will be real, and the refractive index becomes simply $n = \sqrt{\varepsilon_c}$. This is the case with the majority of dielectric thin films that we will be dealing with. However, ε_c can also contain imaginary values. For example, conductive materials will have complex ε_c, and pure dielectrics can also have complex values close to their molecular resonance frequencies. A dielectric like SiO_2 ordinarily has real-valued ε_c but becomes complex at wavelengths close to 9 μm (see Figure 2.1 in Chapter 2). Metals ordinarily have complex ε_c due to their high conductivity σ, but at certain wavelength ranges, ε_c can become real as well. In other words, metals can behave like dielectrics at certain frequency ranges.

1.3 PLANE WAVES

If we assume the medium is uniform and isotropic with no boundaries, then the right-hand side of equation (1.41) will become zero, resulting in

$$\nabla^2 \mathbf{E} + k_0^2 n^2 \mathbf{E} = 0. \tag{1.43}$$

In Cartesian co-ordinates, we can write this as

$$\frac{\partial^2 \mathbf{E}}{\partial z^2} + \frac{\partial^2 \mathbf{E}}{\partial x^2} + \frac{\partial^2 \mathbf{E}}{\partial y^2} + \left(k_x^2 + k_y^2 + k_z^2 \right) \mathbf{E} = 0, \tag{1.44}$$

where we have decomposed the angular wave number into

$$k_0^2 n^2 = k_x^2 + k_y^2 + k_z^2. \tag{1.45}$$

A plane wave maintains a constant direction regardless of propagation distance. In other words, there will be no coupling between the x, y, and z directions. This allows us to separate the x, y, and z directions in equation (1.44) and write

$$\frac{\partial^2 \mathbf{E}}{\partial z^2} + k_z^2 \mathbf{E} = 0 \tag{1.46}$$

$$\frac{\partial^2 \mathbf{E}}{\partial x^2} + k_x^2 \mathbf{E} = 0 \tag{1.47}$$

$$\frac{\partial^2 \mathbf{E}}{\partial y^2} + k_y^2 \mathbf{E} = 0. \tag{1.48}$$

Furthermore, regardless of the actual direction of propagation, we can always re-orient the axes such that the propagation is along z. In that case, only $\frac{\partial^2 \mathbf{E}}{\partial z^2}$ will have a nonzero value, resulting in

$$\frac{\partial^2 \mathbf{E}}{\partial z^2} = -k_z^2 \mathbf{E} \qquad (1.49)$$

$$\frac{\partial^2 \mathbf{E}}{\partial x^2} = 0 \qquad (1.50)$$

$$\frac{\partial^2 \mathbf{E}}{\partial y^2} = 0. \qquad (1.51)$$

The solution of equations (1.49)–(1.51) is simply

$$\mathbf{E} = \mathbf{A} e^{\pm j k_z z}, \qquad (1.52)$$

where $\mathbf{A}$ is the amplitude of the wave. Writing $\mathbf{A}$ as $\hat{\mathbf{e}}A$, where $\hat{\mathbf{e}}$ is the direction of the $\mathbf{A}$ vector (i.e., polarization of the electric field), results in

$$\mathbf{E} = \hat{\mathbf{e}}A e^{\pm j k_z z}. \qquad (1.53)$$

Combining with the assumed time dependence from equation (1.28), the expression for the plane wave becomes

$$\mathbf{E} = \hat{\mathbf{e}}A e^{j(\omega t \pm k_z z)}. \qquad (1.54)$$

Because of the assumed time dependence of the form $e^{j\omega t}$, the negative sign of k_z will represent a wave traveling along the positive z axis. Positive sign of k_z will represent a wave traveling along the negative z axis.

Additionally, from equation (1.36), the refractive index of a conductive material such as a metal or semiconductor will have a negative imaginary part. Therefore, we should be able to verify that a forward traveling wave will exhibit a decaying field and the backward moving wave will exhibit a decaying field in the negative direction. On the other hand, a refractive index with a positive imaginary part would represent amplification (or optical gain). This is only encountered in lasers and optical amplifiers.

Equation (1.54) represents a wave that propagates along a direction k_z with an angular frequency ω. In addition, since ε is uniform in an isotropic material, from equation (1.24), we can write

$$\nabla \cdot \mathbf{E} = 0, \qquad (1.55)$$

from which we can get

$$\pm jA\mathbf{k} \cdot \hat{\mathbf{e}} = 0. \qquad (1.56)$$

Therefore, we can conclude that $\mathbf{k}$ and $\hat{\mathbf{e}}$ will be orthogonal to each other. In other words, the electric field polarization will be orthogonal to the direction of the plane wave's propagation. In fact, the magnetic field is also transverse to the direction of propagation. As a result, this type of wave is referred to as TEM (transverse electromagnetic) waves.

1.4 POWER FLUX

The power dissipation per unit volume of the material is the product of the field intensity and the current density. This essentially follows from the same definition of electrical power as the product of voltage and current:

$$P = \mathbf{E} \cdot \left(\frac{\partial \mathbf{D}}{\partial t} + \mathbf{J} \right) \tag{1.57}$$

$$= \mathbf{E} \cdot \left(\frac{\partial \varepsilon \mathbf{E}}{\partial t} + \sigma \mathbf{E} \right). \tag{1.58}$$

The first term inside the bracket is due to the displacement current, and the second term is due to the conduction current. We can also write this as

$$P = \mathbf{E} \cdot \mathbf{J}', \tag{1.59}$$

where we have combined the displacement and conduction current into a single term $\mathbf{J}'$. Using the assumed time harmonic form in equation (1.54), we can write an expression for the complex admittance of the medium. This is the inverse of the impedance, expressed as

$$Y = \frac{1}{Z} = \frac{\mathbf{J}'}{\mathbf{E}} = j\varepsilon\omega + \sigma. \tag{1.60}$$

The power becomes

$$P = A^2 \left(j\varepsilon\omega + \sigma \right) e^{j2(\omega t \pm k_z z)}, \tag{1.61}$$

and its root mean square (RMS) value is

$$P_{\text{rms}} = \frac{A^2}{2} \left(j\varepsilon\omega + \sigma \right). \tag{1.62}$$

We can draw parallels between this complex power and an RC or RL (resistor–capacitor or resistor–inductor) electrical circuit containing a resistor and a capacitor (or inductor). The real part of P contributes to the irreversible power losses due to joule heating, and the imaginary part contributes to the reactive power. The latter represents the energy that is repeatedly stored and released by the medium, resulting in the propagation of energy. The loss tangent is defined as the ratio between this lossy reaction and the lossless reaction. That is,

$$\tan \delta = \frac{\text{Re}\{Y\}}{|\text{Im}\{Y\}|}. \tag{1.63}$$

We have taken the magnitude of the lossless reaction because the sign simply indicates whether the reaction is primarily capacitive (positive) or inductive (negative).

The directional power flux $\mathbf{S}$ (power flow per unit area or Poynting vector) can be defined as

$$\nabla \cdot \mathbf{S} = -P, \tag{1.64}$$

which states that the differential amount of power flux crossing a unit volume is equal to the power lost within that volume. For a plane wave propagating along the z axis, the derivative of **S** will be nonzero only along z. Therefore, equation (1.64) can be written as

$$\frac{\partial S_z}{\partial z} = -P \tag{1.65}$$

$$S_z = -\int P dz \tag{1.66}$$

$$= \frac{A^2 Y e^{2j(\omega t \pm k_z z)}}{j2k_z}. \tag{1.67}$$

Substituting for k_0 from equation (1.40), and for n from equation (1.36), results in

$$S_z = \frac{A^2}{2} \sqrt{\frac{\varepsilon_0}{\mu_0}} n e^{2j(\omega t \pm k_z z)}. \tag{1.68}$$

The complex admittance Y can be written in terms of the refractive index using equations (1.33) and (1.60) to get

$$Y = \sigma + j\varepsilon\omega \tag{1.69}$$

$$= \left[\frac{\sigma}{\omega\varepsilon_0} + j\frac{\varepsilon}{\varepsilon_0}\right]\omega\varepsilon_0 \tag{1.70}$$

$$= \left[\frac{\varepsilon}{\varepsilon_0} - j\frac{\sigma}{\omega\varepsilon_0}\right]j\omega\varepsilon_0 \tag{1.71}$$

$$= \varepsilon_c j\omega\varepsilon_0 \tag{1.72}$$

$$= jn^2\omega\varepsilon_0 \tag{1.73}$$

$$= j(m - j\kappa)^2\omega\varepsilon_0 \tag{1.74}$$

$$= \left[2m\kappa + j\left(m^2 - \kappa^2\right)\right]\omega\varepsilon_0. \tag{1.75}$$

From this expression, we can see that the absorption losses of the material are characterized by the factor $2m\kappa$. In other words, both the real and imaginary parts are responsible for the absorption losses. The reactive (capacitive or inductive) portion of the admittance is $m^2 - \kappa^2$. Therefore, we can write the loss tangent in terms of m and κ as

$$\tan\delta = \frac{2m\kappa}{|m^2 - \kappa^2|}. \tag{1.76}$$

The majority of transparent dielectric thin films we will deal in this book will have loss tangents of nearly zero. In some materials, such as semiconductors and metals, the loss tangent can be large. It is particularly worth noting that κ and $\tan\delta$ represent different things. Loss tangent represents the irreversible absorption loss. κ represents the field attenuation factor, which produces a term $e^{-k_0\kappa z}$. In other words, field attenuation does not necessarily imply absorption. This is a subtle but important concept. It implies that it is possible for a material with a large κ to have a low absorption loss. This happens with some metals such as silver and gold, which have m in the range of 0.1–0.5 with κ around 5, resulting in a small loss tangent but a large κ. It is also possible, at least hypothetically, for the refractive index to be dominantly imaginary, i.e., $0 - j\kappa$. In this case, even though the field will decay as $e^{-k_0 \kappa \hat{\mathbf{k}} \cdot \mathbf{r}}$, the absorption loss will be zero. This is analogous to evanescent field decay rather than an absorption field decay. We will discuss these aspects in more detail in Chapter 12.

1.5 ELECTROMAGNETIC WAVES ACROSS DIELECTRIC BOUNDARIES

1.5.1 DERIVATION OF THE BOUNDARY CONDITIONS

Due to the stratified nature of thin film structures, the refractive index will be uniform along the transverse directions (x and y), with discrete transitions only along the z direction. Figure 1.5 shows such a boundary with an electromagnetic wave propagating from a material with permittivity ε_{f1} to a material with a different permittivity ε_{f2}. This boundary can be represented as a Heaviside step function:

$$\varepsilon_c (z) = \varepsilon_{f1} + H(z)\left(\varepsilon_{f2} - \varepsilon_{f1}\right). \tag{1.77}$$

As shown in Figure 1.6, ε_c changes from ε_{f1} to ε_{f2} stepwise at $z = 0$. Since the derivative of the Heaviside function is a dirac delta function, we can write

$$\frac{\partial \varepsilon_c}{\partial z} = \delta(z)\left(\varepsilon_{f2} - \varepsilon_{f1}\right). \tag{1.78}$$

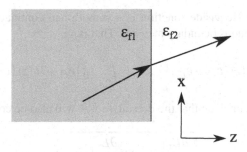

Figure 1.5 Electromagnetic fields across a dielectric boundary.

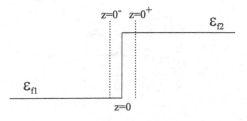

Figure 1.6 Heaviside function representing a discrete transition between films.

Substituting this into the wave equation (1.41) results in

$$\nabla^2 \mathbf{E} + k_0^2 \left[\varepsilon_{f1} + H(z) \left(\varepsilon_{f2} - \varepsilon_{f1} \right) \right] \mathbf{E} = -\nabla \left(\frac{E_z \delta(z) \left(\varepsilon_{f2} - \varepsilon_{f1} \right)}{\varepsilon_c(z)} \right). \quad (1.79)$$

1.5.2 NORMAL INCIDENCE

First, we will consider the case of normal incidence at the boundary, which is much simpler than oblique incidence. We will consider the electric field to be polarized along the x direction and propagating along the z axis, as shown in Figure 1.7. It could also be polarized along the y direction, and the results would be the same. This allows us to write equation (1.79) as

$$\frac{\partial^2 E_x}{\partial z^2} + k_0^2 \left[\varepsilon_{f1} + H(z) \left(\varepsilon_{f2} - \varepsilon_{f1} \right) \right] E_x = -\frac{\partial}{\partial x} \left[\frac{E_z \delta(z) \left(\varepsilon_{f2} - \varepsilon_{f1} \right)}{\varepsilon_c(z)} \right]. \quad (1.80)$$

Since $E_z = 0$ for normal incidence, this simplifies to

$$\frac{\partial^2 E_x}{\partial z^2} + k_0^2 \left[\varepsilon_{f1} + H(z) \left(\varepsilon_{f2} - \varepsilon_{f1} \right) \right] E_x = 0. \quad (1.81)$$

Integrating with respect to z,

$$\frac{\partial E_x}{\partial z} + \int_z k_0^2 \left[\varepsilon_{f1} + H(z) \left(\varepsilon_{f2} - \varepsilon_{f1} \right) \right] E_x dz = 0. \quad (1.82)$$

The integral of the Heaviside function is a smooth and continuous function, even across the discontinuous boundary at $z = 0$. That is,

$$\int_z^{0^-} k_0^2 \left[\varepsilon_{f1} + H(z) \left(\varepsilon_{f2} - \varepsilon_{f1} \right) \right] E_x dz = \int_z^{0^+} k_0^2 \left[\varepsilon_{f1} + H(z) \left(\varepsilon_{f2} - \varepsilon_{f1} \right) \right] E_x dz. \quad (1.83)$$

Therefore, we can conclude that the derivative $\frac{\partial E_x}{\partial z}$ will also be continuous at $z = 0$. That is,

$$\boxed{\left. \frac{\partial E_x}{\partial z} \right|_{z=0^-} = \left. \frac{\partial E_x}{\partial z} \right|_{z=0^+}.} \quad (1.84)$$

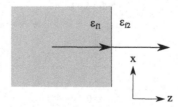

Figure 1.7   Normal incidence across a dielectric boundary.

Similarly, integrating equation (1.82) again, the second integral of the Heaviside function will also be smooth and continuous, resulting in

$$\boxed{E_x|_{z=0^-} = E_x|_{z=0^+} \cdot} \tag{1.85}$$

In summary, under normal incidence, the field and its derivative will be continuous across the dielectric boundary as per equations (1.84) and (1.85).

1.5.3   OBLIQUE INCIDENCE

When the incident field is obliquely incident on the dielectric boundary, the field can be polarized along the y direction, or it can be polarized in the x–z plane. These situations are depicted in Figure 1.8a and b. The first case is referred to as transverse electric (TE) incidence, and the second case is referred to as the transverse magnetic (TM) incidence. This terminology actually comes from waveguides where the direction of propagation of energy is parallel to the dielectric boundary. In the TE case, the propagation of energy (parallel to the dielectric boundary) is transverse to the direction of the electric field. In the TM case, the electric field is not transverse, but the magnetic field is transverse. We will consider each case separately.

1.5.3.1   TE Incidence

In this case, the field will be polarized along the y direction. Because of the oblique incidence, the field derivatives will be nonzero along both the x and z directions. Following the same procedure as with the normal incidence case, we can write equation (1.79) as

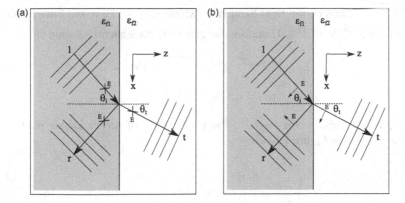

Figure 1.8 Plane waves incident from a material with permittivity ε_{f1} to a second material with permittivity ε_{f2}. The wave is reflected in the incident medium and transmitted into the second medium. (a) Transverse Electric (TE) incidence. (b) Transverse Magnetic (TM) incidence.

$$\frac{\partial^2 E_y}{\partial z^2} + \frac{\partial^2 E_y}{\partial x^2} + k_0^2 \left[\varepsilon_{f1} + H(z) \left(\varepsilon_{f2} - \varepsilon_{f1} \right) \right] E_y = -\frac{\partial}{\partial y} \left[\frac{E_z \delta(z) \left(\varepsilon_{f2} - \varepsilon_{f1} \right)}{\varepsilon_c(z)} \right].$$

(1.86)

Again, the right hand side of this equation will be zero because the field will have no component along the z direction, resulting in

$$\frac{\partial^2 E_y}{\partial z^2} + \frac{\partial^2 E_y}{\partial x^2} + k_0^2 \left[\varepsilon_{f1} + H(z) \left(\varepsilon_{f2} - \varepsilon_{f1} \right) \right] E_y = 0.$$

(1.87)

Since the x and z directions are decoupled for a plane wave, each direction can be separated as

$$\frac{\partial^2 E_y}{\partial x^2} + k_0^2 \varepsilon_{fx} E_y = 0$$

(1.88)

$$\frac{\partial^2 E_y}{\partial z^2} + k_0^2 \left[\varepsilon_{f1} - \varepsilon_{fx} + H(z) \left(\varepsilon_{f2} - \varepsilon_{f1} \right) \right] E_y = 0,$$

(1.89)

where ε_x is the effective dielectric constant for the wave vector along the x direction. We could also define an effective refractive index n_x along the x direction as

$$n_x^2 = \varepsilon_x.$$

(1.90)

The solution of equation (1.88) is a simple plane wave, just like that of our original equation (1.49). Therefore, in the transverse (x) direction, the solution will be of the form $e^{-jk_0\sqrt{\varepsilon_x}x}$. (We are considering only $-x$ in the exponent because it is assumed the incident wave is tilted along the $+x$ direction, and since there are no boundaries in that direction, it will not produce a wave traveling in the $-x$ direction.) In the normal (z) direction, the solution will be governed by equation (1.89). On the left side of the boundary (where $H(z) = 0$), equation (1.89) will have a solution of the form $e^{\pm jk_0\sqrt{\varepsilon_{f1}-\varepsilon_x}\,z}$, and on the right hand side (where $H(z) = 1$), it will have a solution of the form $e^{\pm jk_0\sqrt{\varepsilon_{f2}-\varepsilon_x}\,z}$. Therefore, we can write the general solution as

$$E_y \sim e^{-jk_0\sqrt{\varepsilon_x}\,x} e^{\pm jk_0\sqrt{\varepsilon_{f1}-\varepsilon_x}\,z} \quad \text{for } z < 0$$

(1.91)

$$\sim e^{-jk_0\sqrt{\varepsilon_x}\,x} e^{\pm jk_0\sqrt{\varepsilon_{f2}-\varepsilon_x}\,z} \quad \text{for } z > 0.$$

(1.92)

We can relate these to the angle of the propagation of the incidence plane wave (indicated by θ_i in Figure 1.8) as

$$\sin(\theta_i) = \frac{\sqrt{\varepsilon_x}}{\sqrt{\varepsilon_{f1}}} = \frac{n_x}{n_{f1}}.$$

(1.93)

Similarly, the transmitted angle will be

$$\sin(\theta_t) = \frac{\sqrt{\varepsilon_x}}{\sqrt{\varepsilon_{f2}}} = \frac{n_x}{n_{f2}}.$$

(1.94)

This allows us to write the plane wave expressions (1.91) and (1.92) in terms of refractive indices as

$$E_y \sim e^{-jk_0 n_{f1} \sin(\theta_i) x} e^{\pm jk_0 \sqrt{n_{f1}^2 - n_{f1}^2 \sin^2(\theta_i)}\, z} \quad \text{for } z < 0 \tag{1.95}$$

$$\sim e^{-jk_0 n_{f2} \sin(\theta_t) x} e^{\pm jk_0 \sqrt{n_{f2}^2 - n_{f2}^2 \sin^2(\theta_t)}\, z} \quad \text{for } z > 0. \tag{1.96}$$

How these two expressions behave at the dielectric interface will be determined by the boundary conditions, which can be derived as follows. Integrating equation (1.89) with respect to z results in

$$\frac{\partial E_y}{\partial z} + \int_z k_0^2 \left[\varepsilon_{f1} - \varepsilon_{fx} + H(z)(\varepsilon_{f2} - \varepsilon_{f1}) \right] E_y \, dz = 0. \tag{1.97}$$

Since the integral of the Heaviside function is smooth and continuous, this results in

$$\boxed{\left. \frac{\partial E_y}{\partial z} \right|_{z=0^-} = \left. \frac{\partial E_y}{\partial z} \right|_{z=0^+}.} \tag{1.98}$$

We can integrate equation (1.97) with respect to z again and get

$$\boxed{\left. E_y \right|_{z=0^-} = \left. E_y \right|_{z=0^+}.} \tag{1.99}$$

Additionally, from equation (1.88), the derivative along the x direction will also be continuous across the boundary:

$$\boxed{\left. \frac{\partial E_y}{\partial x} \right|_{z=0^-} = \left. \frac{\partial E_y}{\partial x} \right|_{z=0^+}.} \tag{1.100}$$

Since equation (1.100) has to be satisfied for all x, from equations (1.95) and (1.96), we can conclude that

$$n_{f1} \sin(\theta_i) = n_{f2} \sin(\theta_t), \tag{1.101}$$

which is the same as Snell's law of refraction. Using this in equations (1.95) and (1.96), we can get

$$E_y \sim e^{-jk_0 n_{f1} \sin(\theta_i) x} e^{\pm jk_0 \sqrt{n_{f1}^2 - n_{f1}^2 \sin^2(\theta_i)}\, z} \quad \text{for } z < 0 \tag{1.102}$$

$$\sim e^{-jk_0 n_{f1} \sin(\theta_i) x} e^{\pm jk_0 \sqrt{n_{f2}^2 - n_{f1}^2 \sin^2(\theta_i)}\, z} \quad \text{for } z > 0. \tag{1.103}$$

In summary, for TE incidence, the field and its derivative will be continuous across the boundary. This is identical to the normal incidence case.

1.5.3.2   TM Incidence

TM incidence produces a slightly different result because the field in this case will be polarized in the x–z plane:

$$\mathbf{E} = E_x\hat{\mathbf{x}} + E_z\hat{\mathbf{z}}. \tag{1.104}$$

The wave equation becomes

$$\left(\frac{\partial^2}{\partial z^2} + \frac{\partial^2}{\partial x^2}\right)(E_x\hat{\mathbf{x}} + E_z\hat{\mathbf{z}}) + k_0^2\left[\varepsilon_{f1} + H(z)\left(\varepsilon_{f2} - \varepsilon_{f1}\right)\right](E_x\hat{\mathbf{x}} + E_z\hat{\mathbf{z}}) =$$

$$-\frac{\partial}{\partial x}\left[\frac{E_z\delta(z)\left(\varepsilon_{f2} - \varepsilon_{f1}\right)}{\varepsilon_c(z)}\right]\hat{\mathbf{x}} - \frac{\partial}{\partial z}\left[\frac{E_z\delta(z)\left(\varepsilon_{f2} - \varepsilon_{f1}\right)}{\varepsilon_c(z)}\right]\hat{\mathbf{z}}. \tag{1.105}$$

Each vector component satisfies the wave equation separately, resulting in

$$\frac{\partial^2 E_z}{\partial z^2} + \frac{\partial^2 E_z}{\partial x^2} + k_0^2\left[\varepsilon_{f1} + H(z)\left(\varepsilon_{f2} - \varepsilon_{f1}\right)\right]E_z = -\frac{\partial}{\partial z}\left[\frac{E_z\delta(z)\left(\varepsilon_{f2} - \varepsilon_{f1}\right)}{\varepsilon_c(z)}\right] \tag{1.106}$$

$$\frac{\partial^2 E_x}{\partial z^2} + \frac{\partial^2 E_x}{\partial x^2} + k_0^2\left[\varepsilon_{f1} + H(z)\left(\varepsilon_{f2} - \varepsilon_{f1}\right)\right]E_x = -\frac{\partial}{\partial x}\left[\frac{E_z\delta(z)\left(\varepsilon_{f2} - \varepsilon_{f1}\right)}{\varepsilon_c(z)}\right]. \tag{1.107}$$

1.5.3.2.1   Derivation of Boundary Conditions Using E_z (Equation 1.106)

Let's first consider equation (1.106). As before, we will decouple the x and z directions. However, the component on the right hand side will introduce some differences compared to the TE case. The decoupled equations become

$$\frac{\partial^2 E_z}{\partial x^2} + k_0^2\varepsilon_{fx}E_z = 0 \tag{1.108}$$

$$\frac{\partial^2 E_z}{\partial z^2} + k_0^2\left[\varepsilon_{f1} - \varepsilon_{fx} + H(z)\left(\varepsilon_{f2} - \varepsilon_{f1}\right)\right]E_z = -\frac{\partial}{\partial z}\left[\frac{E_z\delta(z)\left(\varepsilon_{f2} - \varepsilon_{f1}\right)}{\varepsilon_c(z)}\right]. \tag{1.109}$$

As before, the solutions of equations (1.108) and (1.109) will result in a form as expressed in (1.102) and (1.103). The relationship between the fields at the $z = 0$ boundary can be derived as follows.

Integrating equation (1.109) gives

$$\frac{\partial E_z}{\partial z} + \int_z k_0^2\left[\varepsilon_{f1} - \varepsilon_{fx} + H(z)\left(\varepsilon_{f2} - \varepsilon_{f1}\right)\right]E_z dz = -\left[\frac{E_z\delta(z)\left(\varepsilon_{f2} - \varepsilon_{f1}\right)}{\varepsilon_c(z)}\right]. \tag{1.110}$$

The right hand side of equation (1.110) contains a Kronecker delta function. Although this might suggest a discontinuity in $\frac{\partial E_z}{\partial z}$, the delta function has a nonzero value only at exactly $z = 0$. Immediately to the left and right of $z = 0$, its value will be zero. As a result, the value of $\frac{\partial E_z}{\partial z}$ on either side of the boundary will be unaltered by this delta function, resulting in

$$\left.\frac{\partial E_z}{\partial z}\right|_{z=0^-} = \left.\frac{\partial E_z}{\partial z}\right|_{z=0^+}. \tag{1.111}$$

When equation (1.110) is integrated again, the Kronecker delta function will become a Heaviside function:

$$E_z + \int_z \int_z k_0^2 \left[\varepsilon_{f1} - \varepsilon_{fx} + H(z)(\varepsilon_{f2} - \varepsilon_{f1})\right] E_z dz^2 = -\left[\frac{E_z(0) H(z)(\varepsilon_{f2} - \varepsilon_{f1})}{\varepsilon_c(0)}\right]. \tag{1.112}$$

This results in

$$E_z(0^-) = -\int_z \int_z k_0^2 \left[\varepsilon_{f1} - \varepsilon_{fx} + H(z)(\varepsilon_{f2} - \varepsilon_{f1})\right] E_z dz^2 \tag{1.113}$$

$$E_z(0^+) = -\int_z \int_z k_0^2 \left[\varepsilon_{f1} - \varepsilon_{fx} + H(z)(\varepsilon_{f2} - \varepsilon_{f1})\right] E_z dz^2 - \left[\frac{E_z(0)(\varepsilon_{f2} - \varepsilon_{f1})}{\varepsilon_c(0)}\right]. \tag{1.114}$$

We will define the value of ε_c and E_z at $z = 0$ as the average of the two values on either side of the interface:

$$\varepsilon_c(0) = \frac{1}{2}(\varepsilon_{f1} + \varepsilon_{f2}) \tag{1.115}$$

$$E_z(0) = \frac{1}{2}\left[E_z(0^+) + E_z(0^-)\right]. \tag{1.116}$$

Combining equations (1.113)–(1.116) results in

$$E_z(0^-)\varepsilon_{f1} = E_z(0^+)\varepsilon_{f2}. \tag{1.117}$$

In other words, for TM incidence, the normal field components will be discontinuous along z as per equation (1.117), but the field derivative will be continuous as per equation (1.111).

1.5.3.2.2 Derivation of Boundary Conditions Using E_x (Equation 1.107)

Alternatively, we could have also derived the boundary conditions based on the E_x field component. As before, we can separate equation (1.107) into

$$\frac{\partial^2 E_x}{\partial x^2} + k_0^2 \varepsilon_{fx} E_x = 0 \tag{1.118}$$

$$\frac{\partial^2 E_x}{\partial z^2} + k_0^2 \left[\varepsilon_{f1} - \varepsilon_{fx} + H(z)(\varepsilon_{f2} - \varepsilon_{f1})\right] E_x = -\frac{\partial}{\partial x}\left[\frac{E_z \delta(z)(\varepsilon_{f2} - \varepsilon_{f1})}{\varepsilon_c(z)}\right]. \tag{1.119}$$

Again, the solution will have the same form as expressed in equations (1.102) and (1.103).

To get the relationship at the boundary, we can integrate equation (1.119) with respect to z:

$$\frac{\partial E_x}{\partial z} + \int_z k_0^2 \left[\varepsilon_{f1} + H(z) \left(\varepsilon_{f2} - \varepsilon_{f1} \right) \right] E_x dz = -\frac{\partial}{\partial x} \left[\frac{E_z(0) H(z) \left(\varepsilon_{f2} - \varepsilon_{f1} \right)}{\varepsilon_c(0)} \right].$$

$$(1.120)$$

Following the same argument as before, we can get

$$\left. \frac{\partial E_x}{\partial z} \right|_{z=0^-} - \left. \frac{\partial E_x}{\partial z} \right|_{z=0^+} = \frac{\partial}{\partial x} \left[\frac{E_z(0) \left(\varepsilon_{f2} - \varepsilon_{f1} \right)}{\varepsilon_c(0)} \right]. \qquad (1.121)$$

Using the same definitions of $\varepsilon_c(0)$ and $E_z(0)$ from equations (1.115) and (1.116),

$$\left. \frac{\partial E_x}{\partial z} \right|_{z=0^-} - \left. \frac{\partial E_x}{\partial z} \right|_{z=0^+} = \frac{\partial}{\partial x} \left[\frac{\left(E_z(0^-) + E_z(0^+) \right) \left(\varepsilon_{f2} - \varepsilon_{f1} \right)}{\left(\varepsilon_{f2} + \varepsilon_{f1} \right)} \right]. \qquad (1.122)$$

From equation (1.118), we can conclude that E_x and its transverse derivative will be equal everywhere, resulting in

$$\boxed{E_x|_{z=0^-} = E_x|_{z=0^+}} \qquad (1.123)$$

$$\left. \frac{\partial E_x}{\partial x} \right|_{z=0^-} = \left. \frac{\partial E_x}{\partial x} \right|_{z=0^+} = \frac{\partial E_x}{\partial x}. \qquad (1.124)$$

Substituting these into equation (1.121) and utilizing equation (1.117) results in

$$\left. \frac{\partial E_x}{\partial z} \right|_{z=0^-} - \frac{\varepsilon_{f2}}{\left(\varepsilon_{f1} + \varepsilon_{f2} \right)} \frac{\partial E_z(0^-)}{\partial x} = \left. \frac{\partial E_x}{\partial z} \right|_{z=0^+} - \frac{\varepsilon_{f1}}{\left(\varepsilon_{f2} + \varepsilon_{f1} \right)} \frac{\partial E_z(0^+)}{\partial x}. \qquad (1.125)$$

Since equation (1.117) can also be expressed as

$$\varepsilon_{f1} \frac{\partial E_z(0^-)}{\partial x} = \varepsilon_{f2} \frac{\partial E_z(0^+)}{\partial x}, \qquad (1.126)$$

we can subtract equation (1.126) from (1.125) and simplify it, resulting in

$$\left. \frac{\partial E_x}{\partial z} \right|_{z=0^-} - \frac{\partial E_z(0^-)}{\partial x} = \left. \frac{\partial E_x}{\partial z} \right|_{z=0^+} - \frac{\partial E_z(0^+)}{\partial x}. \qquad (1.127)$$

Both the left and right sides of this equation are actually $\nabla \times \mathbf{E}$. Therefore, it is essentially the same as $\nabla \cdot \mathbf{B} = 0$ (equation 1.3).

Additionally, we can express $E_z(0^-)$ and $E_z(0^+)$ in terms of $E_x(0^-)$ and $E_x(0^+)$ as

$$E_z\left(0^-\right) = E_x\left(0^-\right)\tan\left(\theta_i\right) \tag{1.128}$$

$$E_z\left(0^+\right) = E_x\left(0^+\right)\tan\left(\theta_t\right). \tag{1.129}$$

This results in the following boundary condition:

$$\boxed{\left.\frac{\partial E_x}{\partial z}\right|_{z=0^-} - \tan\left(\theta_i\right)\frac{\partial E_x}{\partial x} = \left.\frac{\partial E_x}{\partial z}\right|_{z=0^+} - \tan\left(\theta_t\right)\frac{\partial E_x}{\partial x}.} \tag{1.130}$$

In other words, the tangential field component E_x will be continuous as per equation (1.124), and its derivative will be determined by equation (1.130).

Later, in Chapter 4 we will apply all of these boundary conditions to implement the transfer matrix method.

1.6 PROBLEMS

1. From equation (1.54), show that the phase velocity of the wave is $\frac{1}{n\sqrt{\mu\varepsilon_0}}$.
2. In_2O_3–SnO_2 (ITO) has a refractive index of $1.89 - j0.0023$ at a wavelength of 600 nm. How much would the power flux S_z decline through a 100 nm thick film?
3. If the complex refractive index of a thin film material is $0 - j\kappa$ (zero real part), show that the absorption in the material will be zero. Even with zero absorption, the power flux S will decay as $e^{-2k_0\kappa z}$. Explain how it is possible to have attenuation with no absorption.
4. Show that a metal with a complex refractive index $m - j\kappa$ with $m = \kappa$ (i.e., equal real and imaginary parts) will have the largest possible loss tangent.
5. Although we did not explicitly point this out, total internal reflection is automatically accounted for in the derivations presented in this chapter. Identify which equations would change when total internal reflection occurs.

FURTHER READING

Hummel, R. E. & Guenther, K. H. *Handbook of Optical Properties: Thin Films for Optical Coatings*, Vol. 1, (CRC Press, Boca Raton, FL, 1995). ISBN: 0819462187.

Kaiser, N. and Pulker, H. K. (eds.) *Optical Interference Coatings (Springer Series in Optical Sciences)*, (Springer, Berlin, 2003). ISBN: 3540003649.

Macleod, H. A. *Thin-Film Optical Filters (Series in Optics and Optoelectronics)*, (CRC Press, Boca Raton, FL, 2017). ISBN: 1138198242.

Rancourt, J. D. *Optical Thin Films: User Handbook (SPIE Press Monograph Vol. PM37)*, (SPIE Publications, Bellingham, WA, 1996). ISBN: 0819465097.

Willey, R. R. *Field Guide to Optical Thin Films (SPIE Vol. FG07)*, (SPIE Publications, Bellingham, WA, 2006). ISBN: 0819462187.

2 Optical Thin Film Materials

2.1 PROPERTIES OF OPTICAL THIN FILM MATERIALS

A number of factors need to be considered when selecting materials for use in an optical thin film design. These include optical transparency, refractive index and dispersion (i.e., change in refractive index with wavelength), chemical stability, environmental stability, film stress, ease of deposition, and thickness control. Other factors such as scratch resistance, operating temperatures, and optical damage threshold may also be important considerations in specific applications.

The majority of thin film designs rely on alternating layers of high-refractive-index and low-refractive-index films (the reason for this will be discussed in more detail in Chapter 6). In the visible range, the choice of low-index materials is often between magnesium fluoride (MgF_2), silicon dioxide (SiO_2), and calcium fluoride (CaF_2). Of these, MgF_2 has the lowest refractive index of about 1.38 at a wavelength of 550 nm. Among high-index materials, titanium dioxide (TiO_2) has the highest refractive index of nearly 2.6 at 550 nm. Although other elements such as Si and Ge exhibit even higher refractive indices, they exhibit heavy absorption in the visible spectrum that makes them unsuitable for application in the visible spectrum.

In the infrared spectrum, there is a greater choice of materials, especially for high-index films. For wavelengths longer than 1.1 μm, silicon becomes highly transparent, with a refractive index of 3.6. For wavelengths longer than 1.8 μm, germanium can also be used with an even higher refractive index of 4.05. Both of these materials are popular in infrared applications.

Table 2.1 shows spectral ranges and their nomenclatures most commonly used in optics.

The optical dispersion properties of thin films can be found in published research literature. However, some caution is warranted because there is significant variability

Table 2.1
Commonly Used Divisions of the Optical Spectrum

Band	Wavelengths
Ultraviolet (UV)	200–400 nm
Visible	400–750 nm
Near-infrared (NIR)	750–1,100 nm
Short-wave infrared (SWIR)	1.1–3.0 μm
Mid-wave infrared (MWIR)	3.0–8.0 μm
Long-wave infrared (LWIR)	8.0–15 μm

in the data depending on the type of deposition, temperatures used during growth, substrate type, annealing conditions and even the method used to measure the refractive index. Additionally, the data reported in one article is unlikely to span the entire the spectral range of interest. Hence, this data must be pieced together from multiple sources, many of which may not exactly line up with one another. Compiled archives of refractive index databases are particularly useful, most notably [1–4]. These are widely used by optical thin film scientists as a starting point. It should be emphasized that because the optical properties depend on the deposition conditions, ultimately each researcher must evaluate the films produced from their own tools and not entirely rely on these database values.

2.2   DIELECTRIC THIN FILM MATERIALS

Dielectric films can be broadly grouped based on their chemical composition. The most common groups are oxides, nitrides, fluorides, and sulfides. Less common ones are borates, chlorides, phosphides, etc. Materials within each group share some common properties as well as synthesis methods. Some of the common optical thin film materials and their properties are discussed below.

2.2.1   OXIDES

2.2.1.1   SiO$_2$

Silicon dioxide (SiO$_2$) is the most ubiquitous optical material known simply as "glass," and is used in the manufacture of optical components such as lenses, mirrors, and windows (Figure 2.1). Ordinary glass contains a number of additives such

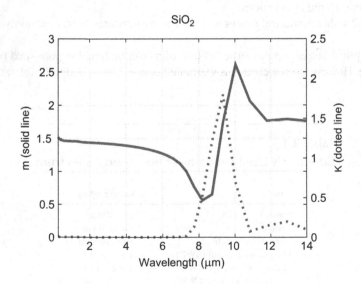

Figure 2.1   Refractive index $n = (m - j\kappa)$ of SiO$_2$. (Based on data from Ref. [2].)

as boron oxide (borosilicate glass), sodium and calcium oxides (soda lime glass), and phosphorous oxide (phosphate glass). These additives achieve different objectives, one of which is a reduction in its melting temperature to make it easier to form the glass into different shapes. In its pure form, SiO_2 is known as silica, and in its crystalline form, it is known as quartz. Thin films of SiO_2 can be produced using physical vapor deposition (PVD) methods, directly from SiO_2 or by oxidizing Si. Chemical vapor deposition with SiH_4 (silane) as a precursor gas is also widely used to make SiO_2. The refractive index of SiO_2 is about 1.48 in the visible to SWIR bands with a fairly low dispersion. It starts to exhibit some absorption in the MWIR which makes it unsuitable as a substrate material, but it can still be used in thin-film form. In the LWIR band, it has significant absorption which makes it unsuitable even as a thin film material.

2.2.1.2 TiO₂

Titanium dioxide, also known as titania, is another widely used optical thin-film material, mostly because it has the highest refractive index in the visible spectrum and has a broad spectral transparency (Figure 2.2). As a result, it is widely used as the high-index film in multilayer thin-film structures. In its crystalline form, it is known as rutile. It is most commonly deposited from a titanium source by reactive physical vapor deposition methods. There are many different stable oxidation states of titanium oxide, each with a different refractive index. Achieving the correct stoichiometry requires careful control of the deposition conditions such as substrate temperature, oxygen pressure, and deposition rate. Besides its use in thin films, TiO_2 is also widely used as a pigment in white paints, plastics, and even in toothpastes.

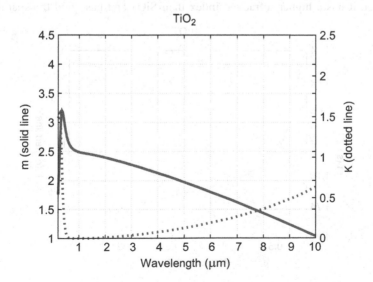

Figure 2.2 Refractive index $n = (m - j\kappa)$ of TiO_2. (Based on data from Refs. [1,4].)

2.2.1.3 Al_2O_3

Aluminum oxide, also known as alumina, is a hard film with a wide optical transparency (Figure 2.3). Its crystalline form is known as sapphire, which is a substrate used for growing several types of semiconductor epitaxial films. The production of sapphire substrate has grown in recent years because it is one of the few substrates on which GaN epitaxial films could be grown (which is used in the manufacture of white LEDs). The optical absorption of alumina is low from UV to SWIR, and its dispersion is also fairly flat. Although its absorption is lower than SiO_2 in the MWIR, it is still significant enough to make alumina unusable as a substrate in this wavelength. Al_2O_3 can be deposited by reactive physical vapor deposition methods from an aluminum source. Because it has a low erosion rate when irradiated with ions, it is commonly used to make fixtures in plasma chambers. However, the low erosion rate also implies a low sputter deposition rate, so it cannot be easily sputtered. However, it can be directly evaporated from an Al_2O_3 source.

2.2.1.4 Ta_2O_5

Tantalum pentoxide, or tantala, is a high-index low-loss material (Figure 2.4). In addition to its use as an optical thin film, it is also used in tantalum capacitors and as a dielectric layer in metal–oxide–semiconductor (MOS) transistors. Optical thin films of Ta_2O_5 are most commonly deposited by physical vapor deposition methods.

2.2.1.5 SiO

Silicon monoxide is a variant of silicon dioxide (Figure 2.5). It has a brown tint due to its absorption in the visible spectrum, which makes it unsuitable in this range. However, it has a higher refractive index than SiO_2 and has good transparency in

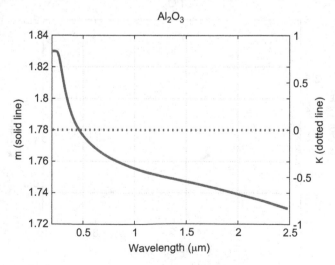

Figure 2.3 Refractive index $n = (m - j\kappa)$ of Al_2O_3. (Based on data from Ref. [3].)

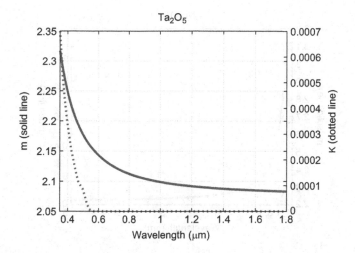

Figure 2.4 Refractive index $(m - j\kappa)$ of Ta_2O_5. (Based on data from Ref. [2].)

the SWIR and MWIR. Its refractive index makes it an ideal antireflection film for silicon, especially in the MWIR spectrum. It is environmentally stable and is easily deposited by thermal evaporation.

2.2.1.6 Nb₂O₅

Niobium pentoxide is a high-index material with high transparency across the visible and SWIR bands (Figure 2.6). It is a very stable material and can be deposited by reactive PVD from metallic niobium or directly from a Nb_2O_5 source.

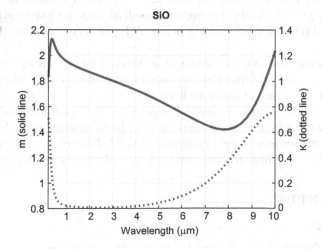

Figure 2.5 Refractive index $n = (m - j\kappa)$ of SiO. (Based on data from Refs. [1,4].)

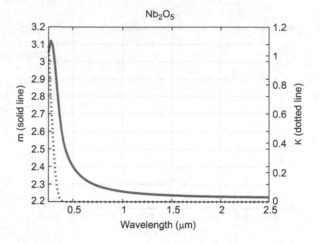

Figure 2.6 Refractive index $n = (m - j\kappa)$ of Nb$_2$O$_5$. (Based on data from Ref. [2].)

2.2.1.7 Other Oxide Films

Although the oxides listed above are few of the most common optical thin-film materials, a number of other oxides are also used in the design and manufacture of optical thin films. Tungsten oxide (WO$_3$) is a high-index material with low losses in the visible to SWIR, and it is very easily deposited using thermal evaporation. It is also an electrochromic film – its refractive index can be dramatically altered by the infusion of protons or lithium ions.

Indium tin oxide, abbreviated as ITO, is a mixture of In$_2$O$_3$ and SnO$_2$ (typically 90% and 10%) with a refractive index of about 1.8–2.0. This is widely used for making transparent conductors. It has excellent transparency in the visible range and a resistivity of about 1 mΩ-cm. Although its resistivity is about three orders of magnitude larger than metals like copper, it is still adequate for low-switching-speed applications. ITO is currently used in nearly all liquid crystal displays (LCDs) and touch screens.

Vanadium dioxide (VO$_2$) is another oxide that has a large refractive index. It is rarely used as an optical thin film because its deposition is quite difficult. However, it belongs to a class of materials known as phase change materials (PCMs) that undergo a reversible physical transformation. VO$_2$ switches from a dielectric to a metallic state at a transition temperature of 68°C. Its multivalent oxide, VO$_x$, is used as a temperature sensing element in uncooled microbolometer thermal imagers. The design of optical thin-film structures using PCMs is discussed in Chapter 13.

2.2.2 FLUORIDES

2.2.2.1 MgF$_2$

Magnesium fluoride is widely used in optical thin films because it has one of the lowest refractive indices (Figure 2.7). It's refractive index is about 1.38 in the visible spectrum declining to about 1.3 in the infrared. MgF_2 can be sputtered or evaporated from a MgF_2 source. The imaginary part of refractive index is fairly small making it suitable as a thin film in the visible spectrum to MWIR. MgF_2 can also be deposited from metallic magnesium by reactive PVD using a fluorine-containing gas, such as CF_4 or SF_6.

2.2.2.2   CaF$_2$

Calcium fluoride also has a low refractive index, although somewhat higher than MgF_2. It is in the range of 1.4 in the visible spectrum (Figure 2.8). Most importantly, CaF_2 is very transparent in the deep-UV wavelengths where even fused silica becomes opaque. As a result, it is used to make optical elements in deep-UV lithography tools.

2.2.3   NITRIDES

2.2.3.1   Si$_3$N$_4$

Silicon nitride is a popular thin film in the microfabrication area, especially in micro-electromechanical systems (MEMs). It is strong, chemically inert, and has a moderately high refractive index with low optical losses (Figure 2.9). It is also commonly used as an antireflection film for silicon, in applications such as photodetectors and image sensors. Si_3N_4 can be deposited using any of the PVD methods or chemical vapor deposition (CVD) methods.

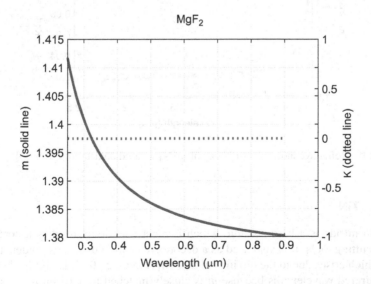

Figure 2.7   Refractive index $n = (m - j\kappa)$ of MgF_2. (Based on data from Ref. [3].)

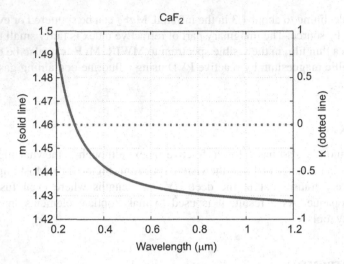

Figure 2.8 Refractive index $n = (m - j\kappa)$ of CaF_2. (Based on data from Ref. [3].)

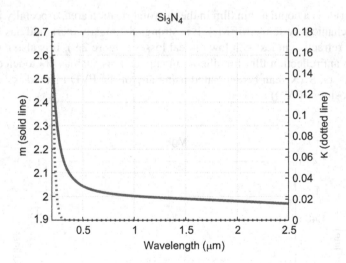

Figure 2.9 Refractive index $n = (m - j\kappa)$ of Si_3N_4. (Based on data from Ref. [3].)

2.2.3.2   TiN

Titanium nitride is a hard ceramic which is most commonly used as a coating to harden cutting tools. It is also used as a decorative coating due to its golden appearance, which arises due to the dip in m at 500 nm (Figure 2.10). It is highly absorbing at all infrared wavelengths because m is closely matched to κ (resulting in a large loss tangent). It can be deposited by PVD or CVD methods.

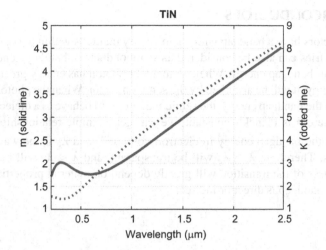

Figure 2.10 Refractive index $n = (m - j\kappa)$ of TiN. (Based on data from Refs. [1,4].)

2.2.4 SULFIDES

2.2.4.1 ZnS

Zinc sulfide is a high-index material, and it is one of the few materials that are transparent from the visible spectrum to the LWIR. Hence it is a very popular material in infrared optics. In the visible and NIR spectrum, it has a refractive index of 2.3 (Figure 2.11). It also starts absorbing around 400 nm, which gives it the yellow hue. ZnS is most commonly deposited by PVD methods such as sputtering or evaporation directly from a ZnS source.

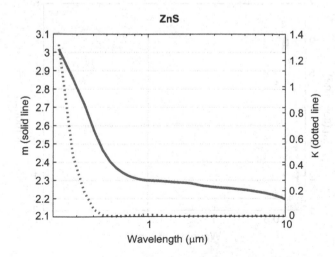

Figure 2.11 Refractive index $n = (m - j\kappa)$ of ZnS. (Based on data from Ref. [2].)

2.3 SEMICONDUCTORS

Semiconductors have a bandgap with a gap energy that falls within the optical range. Many dielectrics can also be considered as semiconductors, but they generally have a much larger bandgap energy. When an incident photon has energy greater than the bandgap energy, it will be absorbed by the semiconductor. When the photon energy is smaller than the bandgap energy, the semiconductor will behave as a dielectric. Using $E = h\nu$, where h is Planck's constant, we can get a simple relationship $\lambda_g = \frac{1.24}{E_g}$ where E_g is the bandgap energy in electron volts (eV) and λ_g is the wavelength in micrometers. Therefore, $\lambda > \lambda_g$ will be transparent, and $\lambda < \lambda_g$ will be absorbed. The abruptness of the transition will greatly depend on material properties, e.g., on whether the bandgap is direct or indirect.

2.3.1 Si

Silicon is the most studied semiconductor material because of its wide use in electronics (Figure 2.12). It has an indirect bandgap of 1.1 eV ($\lambda_g = 1.1\mu m$). As a result, it is opaque in the visible spectrum but becomes transparent for wavelengths longer than 1.0 μm and continues well into the LWIR. Electronic applications require the silicon to be crystalline, but in optical thin-film applications, it can be amorphous or polycrystalline. It is a high-index material with a refractive index of 3.5 in the infrared. It can be deposited using a number of different PVD and CVD methods. It is highly reactive, so it requires an ultrapure environment to prevent oxidation during deposition.

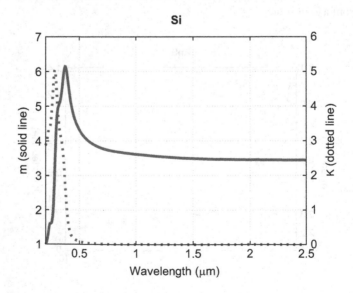

Figure 2.12 Refractive index $n = (m - j\kappa)$ of Si. (Based on data from Ref. [3].)

2.3.2   Ge

Germanium has properties very similar to silicon. It has an indirect bandgap energy of 0.66 eV ($\lambda_g = 1.87\mu$m) (Figure 2.13). It is transparent for wavelengths longer than 1.9 μm. It has one of the highest refractive indices of all thin-film materials, with an index of 4.05 in the infrared.

2.3.3   CdS

Cadmium sulfide (CdS) is a high-index semiconducting material with a bandgap of 2.42 eV ($\lambda_g = 512$ nm) (Figure 2.14). Thin films of CdS are commonly produced by sputter deposition. It is used as the active film in photocells (photoconductors) and in thin-film solar cells.

2.4   METALS

2.4.1   Ag

Silver has a special place in optics because it is partially transparent in the visible spectrum. The small value for m makes the loss tangent of silver one of the smallest of all metals (Figure 2.15). Combined with the relatively small κ, the optical fields are able to penetrate deeper into the metal with little losses. Silver is widely used in transparent metal designs and architectural windows. While it is transparent in the visible spectrum, it becomes highly reflective in the infrared spectrum, so it can serve as energy-efficient windows. It is fairly easy to deposit silver, but achieving ultrathin

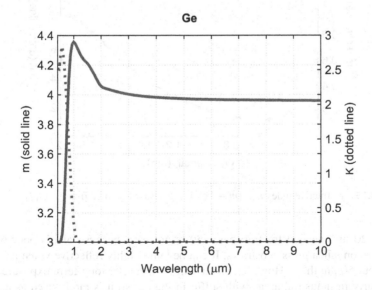

Figure 2.13   Refractive index $n = (m - j\kappa)$ of Ge. (Based on data from Ref. [2].)

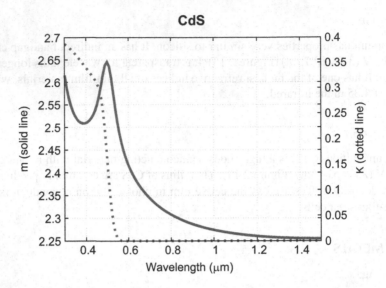

Figure 2.14 Refractive index $n = (m - j\kappa)$ of CdS. (Based on data from Ref. [2].)

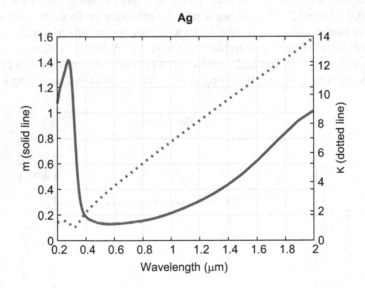

Figure 2.15 Refractive index $n = (m - j\kappa)$ of Ag. (Based on data from Ref. [3].)

films (<20 nm, needed for transparency) is not easy because of it's poor wetting properties on silica glass substrates. It can become highly reflective when it is more than about 50 nm thick. However, silver will tarnish after long-term exposure to air, particularly in industrial areas with sulfur in the air, so it is rarely used as a mirror coating by itself.

2.4.2 Al

Aluminum is highly reflective, and it is widely used as reflective coating in mirrors. However, it has a weak resonance at 800 nm, which results in a dip in reflection (Figure 2.16). Therefore, its application is limited to reflective coatings in the UV and visible spectra. Aluminum also oxidizes quickly, but the oxide forms a strong passivation layer that protects it from further oxidation.

2.4.3 Au

Gold is widely used in electronics and in optics because it is environmentally stable. The color of gold arises due to its low reflectance in the blue and an increase in reflectivity at 550 nm (Figure 2.17). However, above 1 μm, it is an excellent infrared reflector, so it is often used as a calibration standard in many infrared systems.

2.4.4 Cu

Copper can be partially transparent in the visible spectrum when it is ultrathin. It also has reasonably low loss tangents and can be easily deposited (Figure 2.18).

2.4.5 Cr

Chromium is not a metal of particular interest in optics, but it is often used as an underlayer to improve the adhesion of many metals, especially gold and silver. Therefore, it's optical properties become relevant for many optical thin-film studies. Additionally, its loss tangent is extremely high, so it is useful as a broadband optical absorber (Figure 2.19). Figure 2.20 shows the dispersion properties of platinum.

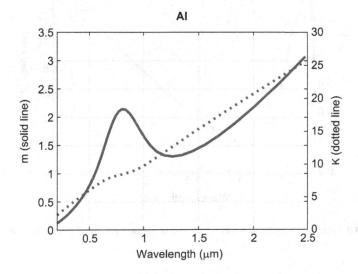

Figure 2.16 Refractive index $n = (m - j\kappa)$ of Al. (Based on data from Ref. [3].)

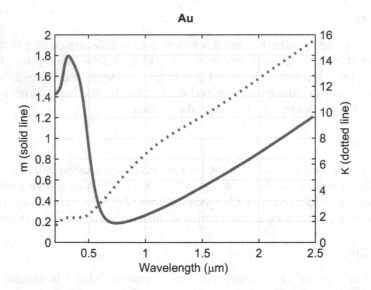

Figure 2.17 Refractive index $n = (m - j\kappa)$ of Au. (Based on data from Ref. [3].)

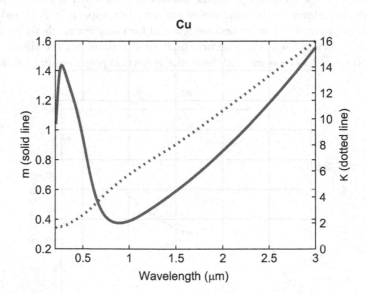

Figure 2.18 Refractive index $n = (m - j\kappa)$ of Cu. (Based on data from Refs. [1,4].)

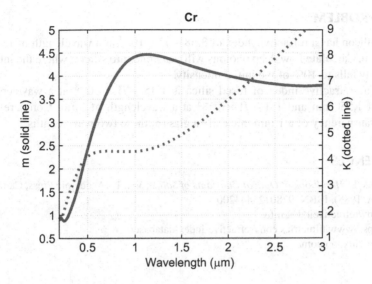

Figure 2.19 Refractive index $n = (m - j\kappa)$ of Cr. (Based on data from Refs. [1,4].)

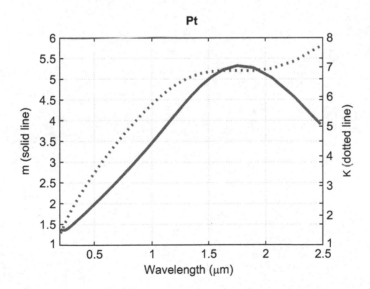

Figure 2.20 Refractive index $n = (m - j\kappa)$ of Pt. (Based on data from Ref. [3].)

2.5 PROBLEMS

1. Silicon has a refractive index of $3.58 - j7 \times 10^{-4}$ at a wavelength of $1,064$ nm. Calculate how deep photons will penetrate into silicon where the intensity falls to 10% of the initial intensity.
2. The refractive index of fused silica is $1.45 - j1 \times 10^{-8}$ at a wavelength of 1,550 nm and $1.3 - j1 \times 10^{-3}$ at a wavelength of 5μm. Compare the transparency of a 1 mm thick silica glass at these two wavelengths.

REFERENCES

1. Palik, E. *Handbook of Optical Constants of Solids*, Vol. 1. (Academic Press, Cambridge, MA, 1985). ISBN: 9780125444200.
2. http://refractiveindex.info.
3. https://www.filmetrics.com/refractive-index-database/.
4. http://luxpop.com/.

3 Single-Layer Antireflection Theory

3.1 REFLECTION FROM A SINGLE DIELECTRIC INTERFACE

We can derive the reflection coefficient at normal incidence at a dielectric interface by considering a substrate with a refractive index of n_s and an outside medium (normally air) with a refractive index n_a, as shown in Figure 3.1. We will assume a plane wave normally incident at the interface, producing a reflected and a transmitted plane wave. We will assume the amplitude of the incident wave to be 1.0, which enables us to interpret the amplitudes of the other two waves as the field reflection coefficient r and the field transmission coefficient t. By ensuring the continuity of the electric field and its derivative at the interface, we can derive the expressions for reflection and transmission.

For the field continuity, we can use equation (1.85) from Chapter 1, resulting in

$$e^{-jk_0 n_a z_1} + re^{+jk_0 n_a z_1} = te^{-jk_0 n_s z_1}. \tag{3.1}$$

For the field derivative, we can use equation (1.84) from Chapter 1,

$$n_a e^{-jk_0 n_a z_1} - r n_a e^{+jk_0 n_a z_1} = n_s t e^{-jk_0 n_s z_1}. \tag{3.2}$$

We can simplify the above two expressions by locating the origin at the dielectric interface. Setting $z_1 = 0$, equations (3.1) and (3.2) become

$$1 + r = t \tag{3.3}$$

$$n_a - r n_a = n_s t. \tag{3.4}$$

Eliminating t from equations (3.3) and (3.4) results in

$$r = \frac{n_a - n_s}{n_a + n_s}. \tag{3.5}$$

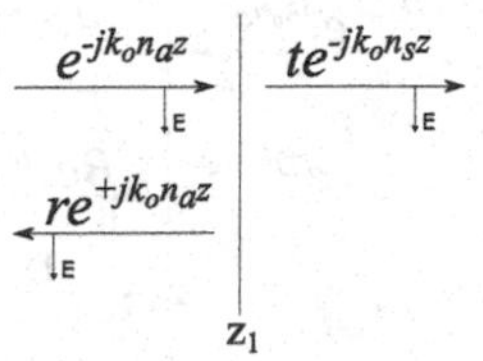

Figure 3.1 Reflection and transmission at normal incidence from a single dielectric interface.

39

This is the well-known Fresnel reflection coefficient for the field amplitudes at normal incidence. The power reflection coefficient R can be computed by considering the ratio of the incident and reflected directional power flux

$$R = \frac{|S_r|}{|S_i|}. \tag{3.6}$$

Using equation (1.68) from Chapter 1,

$$|S_i| = \frac{|A_i|^2}{2}\sqrt{\frac{\varepsilon_0}{\mu_0}}n_a \tag{3.7}$$

$$|S_r| = \frac{|A_r|^2}{2}\sqrt{\frac{\varepsilon_0}{\mu_0}}n_a. \tag{3.8}$$

Therefore,

$$R = \frac{|S_r|}{|S_i|} = \frac{|A_r|^2}{|A_i|^2} = \frac{|r|^2}{1} = |r|^2. \tag{3.9}$$

Therefore, the expression for reflection finally becomes

$$R = \left|\frac{n_a - n_s}{n_a + n_s}\right|^2. \tag{3.10}$$

As an example, for silica glass with a refractive index of 1.48, the air/glass interface will have a reflection of 3.7%. For silicon, which has a refractive index of about 3.5 in the SWIR wavelengths, the reflection will be about 31%.

3.2 SINGLE-FILM ANTIREFLECTION

The concept of antireflection is based on introducing a thin film on the substrate which will produce a reflection equal in amplitude but opposite in phase to cancel out the substrate reflection. This is illustrated in Figure 3.2a. To produce this phase reversal, the film thickness has to be equivalent to a phase shift of $\pi/2$. The round trip phase, therefore, will be π. Furthermore, the refractive index of the film should be such that the amplitudes of the reflections from the top and bottom interfaces are equal. This places two limits on the film: (1) it's refractive index and (2) it's thickness.

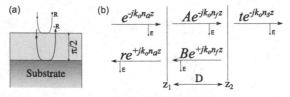

Figure 3.2 Single-film antireflection. (a) Basic concept of single film antireflection. (b) Reflection from a film of thickness D on a substrate.

In order to mathematically evaluate this required condition for antireflection, we need to get an expression for the overall reflection. This can be derived similar to the approach presented in Section 3.1. Referring to Figure 3.2b,

$$e^{-jk_0 n_a z_1} + re^{+jk_0 n_a z_1} = Ae^{-jk_0 n_f z_1} + Be^{+jk_0 n_f z_1} \tag{3.11}$$

$$n_a e^{-jk_0 n_a z_1} - rn_a e^{+jk_0 n_a z_1} = n_f Ae^{-jk_0 n_f z_1} - n_f Be^{+jk_0 n_f z_1}. \tag{3.12}$$

At the $z = z_2$, the two expressions become

$$Ae^{-jk_0 n_f z_2} + Be^{+jk_0 n_f z_2} = te^{-jk_0 n_s z_2} \tag{3.13}$$

$$n_f Ae^{-jk_0 n_f z_2} - n_f Be^{+jk_0 n_f z_2} = n_s te^{-jk_0 n_s z_2}. \tag{3.14}$$

The above four equations can be solved for the four unknowns A, B, r, and t. It involves some messy algebra, but the solution for r works out to be

$$r = \frac{(n_a - n_r)}{(n_a + n_r)}, \tag{3.15}$$

where n_r is defined as

$$n_r = n_f \frac{(n_s + n_f) + (n_s - n_f) e^{-j2\theta}}{(n_s + n_f) - (n_s - n_f) e^{-j2\theta}}, \tag{3.16}$$

and θ is the phase thickness of the film given by

$$\theta = k_0 n_f (z_2 - z_1) = k_0 n_f D. \tag{3.17}$$

Comparing equation (3.15) with equation (3.5), we can see that the substrate index n_s has essentially been replaced with n_r. Therefore, we can consider n_r as the *effective reflectance index* of the substrate due to the presence of the thin film.

We can obtain a perfect antireflection condition if n_r can be somehow made equal to n_a. This would make the substrate appear as if it had an index of n_a. Since n_a is real, n_r also has to be real. Therefore, the exponential terms in equation (3.16) must also be real, resulting in

$$e^{-j2\theta} = \pm 1 \tag{3.18}$$

or

$$\theta = N\frac{\pi}{2}, \tag{3.19}$$

where N is an integer. First, let's consider only odd values of N (or $e^{-j2\theta} = -1$). This would make n_r in equation (3.16) to become

$$n_r = \frac{n_f^2}{n_s}. \tag{3.20}$$

Substituting $n_r = n_a$, the required film index for antireflection becomes

$$n_f = \sqrt{n_a n_s}. \tag{3.21}$$

Then, from equation (3.17), the required film thickness becomes

$$k_0 n_f D = N \frac{\pi}{2} \tag{3.22}$$

$$D = N \frac{\lambda}{4 n_f}. \tag{3.23}$$

In other words, the film has to be an odd integer multiple of a quarter-wave thickness, and the refractive index of the film must be equal to $\sqrt{n_a n_s}$.

On the other hand, even values of N (or $e^{-j2\theta} = +1$ in equation (3.18)) would make $n_r = n_s$. This is not a useful condition because the effective reflectance index will be unchanged from its original substrate value regardless of the film index. This should not be a surprising result because the round trip phase will be 2π when the phase thickness of the film is π. This would make the film behave exactly the same as a zero-thickness (nonexistent) film.

For example, consider a silica glass substrate with a refractive index of $n_s = 1.48$. If we need antireflection at a wavelength of $\lambda = 550$ nm, the required film must have a refractive index of $n_f = \sqrt{1.48} = 1.217$, and the film thickness has to be $\frac{\lambda}{4n_f} = 113$ nm. If the film thickness is 226 nm (which is equivalent to a 2π phase shift), the reflection at 550 nm will be back to its original 3.7%. Following the same argument, we can see that an $N = 3$ (three quarter waves thick) will result in the same antireflection condition as when $N = 1$ (single quarter wave thick).

As a second example, for producing a single-layer antireflection for a silicon substrate at a wavelength of 1,500 nm, the required film must have a refractive index of $n_f = \sqrt{3.5} = 1.87$, and the film thickness must be $\frac{\lambda}{4n_f} = 200.5$ nm.

3.3 COMPLEX EFFECTIVE REFLECTANCE INDEX CONTOURS

Instead of limiting our discussion to real values of n_r, we can also consider the entire range of values of n_r described by equation (3.16). This can be visualized by plotting the complex function n_r as a function of θ (or equivalently, film thickness) as shown in Figure 3.3.

We will use the silica glass substrate as the example with a quarter-wave film of index 1.217. The curve describes an arc in the complex plane. When $\theta = 0$ (when there is no film present), the value of n_r will be equal to the substrate index, which in this case is $n_s = 1.48$. As θ increases, n_r becomes complex and eventually approaches $n_r = 1.0$ when $\theta = \frac{\pi}{2}$. This is the antireflection condition. For all intermediate values of θ, the combination of the substrate and film can be treated as having a complex effective reflectance index. The complex value simply represents the phase of the reflection in equation (3.15); it should not be interpreted as representing attenuation or absorption.

We can also examine what happens if θ increases beyond $\frac{\pi}{2}$. This is shown in Figure 3.4. As θ increases beyond $\frac{\pi}{2}$, the curve moves upward in the complex plane and returns back to the starting point of $n_r = n_s$ when $\theta = \pi$. When $\theta = \pi$, the film behaves as if it does not exist at all. This is an interesting (and useful) condition and

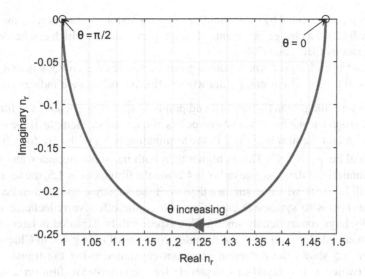

Figure 3.3   Complex contour of n_r as a function of increasing film thickness.

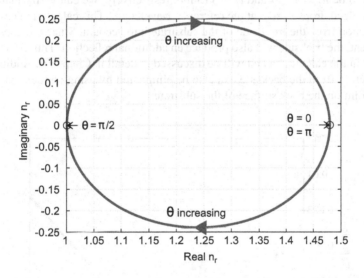

Figure 3.4   Complex contour of n_r for a half-wave thick film.

is known as the absentee film condition. It corresponds to even-integer values of N in equation (3.19) or multiples of half-wave film thicknesses.

An increase in θ can be viewed as arising due to an increase in film thickness at a fixed wavelength or as arising from a declining wavelength (increasing k) for a fixed film thickness. From equation (3.17), we can see that both are equivalent. In a typical application, one would choose a film thickness D that would produce $\theta = \frac{\pi}{2}$ at the desired antireflection wavelength. We define this wavelength as the *reference wavelength*, represented by the symbol λ_0. Any excursions from this reference wavelength will cause a change in the antireflection performance, which can be evaluated from equations (3.15) and (3.16).

Another useful aspect worth pointing out in Figures 3.3 and 3.4 is that the arc from $\theta = 0$ to $\theta = \frac{\pi}{2}$ has end points whose effective reflectance indices are n_s and $\frac{n_f^2}{n_s}$. Therefore, the product of the two end points is n_f^2. In other words, the film index n_f is the *geometric mean* of the two end points of the arc. For example, if the substrate index is 1.5 and the film index is 2.0, the terminating point of the arc for a film with $\theta = \frac{\pi}{2}$ will be $\frac{2^2}{1.5} = 2.67$. This is higher than both the substrate index and the film index. Similarly, if the substrate index is 4.0 and the film index is 1.5, the terminating index will be 0.56, which is smaller than both the substrate and film indices. This aspect allows us to synthesize conditions where the effective reflectance index is artificially large (or artificially small). This aspect will be explored in later chapters. The Python code for producing the complex contours of n_r is given in Chapter 15.

Figure 3.5 shows the reflection spectrum calculated using the transfer matrix method (which is developed in Chapter 4) for a quarter-wave film on a substrate at a reference wavelength of $\lambda_0 = 550$ nm. The film index and film thickness are assumed to be $n_f = 1.217$ and $t_f = 113$ nm, respectively. We can verify that the reflection indeed drops to zero at the reference wavelength. This calculation only takes the reflection from the front side of the substrate into account. Since the substrate is transparent, the backside will also produce an additional reflection. This effect can be included in the calculation and will be discussed in detail in Chapter 4. Additionally, the reflection from the backside can also be eliminated by using an identical antireflection film on the back surface of the substrate.

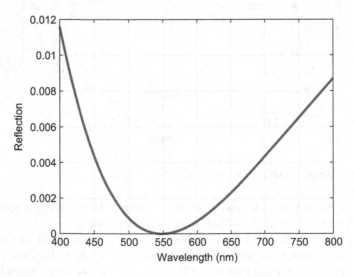

Figure 3.5 Reflection spectrum of a quarter-wave film on a substrate using a reference wavelength of 550 nm.

3.4 LIMITATIONS OF THE EFFECTIVE REFLECTANCE INDEX

A major limitation in the concept of effective reflectance index n_r is that it does not have a direct physical meaning beyond the Fresnel reflection equation (3.15). In other words, it cannot be used in any other context other than for calculating the reflection from thin-film interfaces. It also cannot be used for transmission calculations or resonance effects. Nevertheless, it does have very important applications in thin-film design as we will explore in subsequent chapters.

Additionally, although materials with complex refractive indices will exhibit absorption, a complex n_r should not be interpreted as an absorbing film. The complex value of n_r is simply for representing the phase of the reflected field and not necessarily for absorption or evanescent behavior.

3.5 QUALITY FACTOR

Quality factor, or Q-factor, is often used to describe the strength of resonators. It is the ratio between the energy stored in the resonator and the power loss. Therefore, it may appear strange to bring up Q-factors in the context of antireflection designs. However, an antireflection design is indeed a resonator, albeit a weak one. The minimum reflection (or maximum transmission) occurs when the system is in resonance. In that sense, the process of creating antireflection can be thought of as creating a resonator on a substrate. The Q-factor of the antireflection resonator can be calculated, but it is rarely done. The Q-factor of a single-layer antireflection design is actually very small. This is actually a good thing because it is the small Q-factor (poor resonance) that gives it the broad spectral performance. High Q-factor designs will be have a narrower bandwidth. We will see this in Chapter 5 where increasing the number of layers can result in a narrower bandwidth (and consequently a higher Q-factor).

3.6 NORMALIZED FREQUENCY

Although it is common practice in optics to deal with wavelength rather than the frequency, many mathematical quantities actually scale linearly with frequency instead of wavelength. Therefore, it is also useful to define a normalized frequency

$$G = \frac{f}{f_0}, \tag{3.24}$$

where f_0 is the frequency of the reference wavelength λ_0. In terms of wavelength, the normalized frequency G will be

$$G = \frac{\lambda_0}{\lambda}. \tag{3.25}$$

When $G = 1$, the wavelength will be equal to the reference wavelength. When $G = 2$, the wavelength will be one-half of the reference wavelength (or frequency will be twice the reference frequency). A design that has $\theta = \frac{\pi}{2}$ at $G = 1$ will become

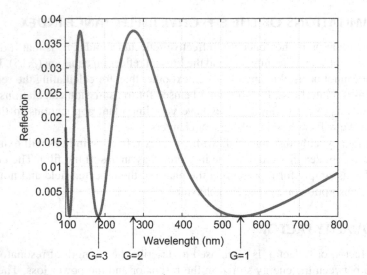

Figure 3.6 Reflection as a function of normalized frequency.

$\theta = \pi$ at $G = 2$. In other words, a quarter-wave antireflection film at $G = 1$ will become a half-wave absentee film at $G = 2$. Therefore, we can expect the film to behave as an antireflection film at odd multiples of G and as an absentee film at even multiples of G.

The 113 nm thick film with index 1.217 for silica glass substrate was designed for $\lambda_0 = 550$ nm. This corresponds to $G = 1$. At $G = 2$ (or $\lambda = 275$ nm), it becomes an absentee film, and the reflection reverts to that of the bare glass substrate of 3.7% (ignoring the reflection from the backside of the substrate). At $G = 3$ (or $\lambda = 183$ nm), the film becomes antireflecting again. This is shown in Figure 3.6.

Nevertheless, there is a practical problem in the above examples with regards to the refractive index of 1.217. It is far too small to be realized with conventional thin-film coating methods. The refractive index of MgF_2, which has the smallest index among the commonly used thin-film materials, is about 1.38. Polymers do exist with lower refractive indices but not as low as 1.217. The only way to solve this problem is by increasing the number of design variables. If we can use more than one film, it can provide more degrees of freedom to allow us to select more realistic materials. This is the one of the reasons for exploring multilayer antireflection coatings.

3.7   PROBLEMS

1. Assuming silicon has a refractive index of 3.5 at a wavelength of 1,550 nm, determine the thickness and refractive index of the required film to produce antireflection at 1,550 nm. Identify which film material is likely to satisfy this requirement.
2. Using the method outlined in Section 3.2, derive equation (3.16).

3. If one were to use MgF_2 as the antireflection film on a silica glass substrate (refractive index of 1.48), calculate the reflection value at the reference wavelength.

4. Assuming aluminum oxide (sapphire) has a refractive index of 1.75 at 1,064 nm, find the appropriate film that will produce antireflection at this wavelength, and plot the effective reflectance index contour in the complex plane up to a half-wave phase thickness.

5. Rutile (TiO_2) has a refractive index of 2.67 at a wavelength of 532 nm. LaF_3 which has an index of 1.60 was used to produce a single-layer antireflection at 532 nm. Subsequently, it was decided that antireflection was required at 1,550 nm and not at 532 nm. The refractive indices of TiO_2 and LaF_3 at 1,550 nm are 2.47 and 1.57, respectively. Calculate the additional thickness of LaF_3 that needs to be deposited to produce antireflection at 1,550 nm and the minimum reflection that can be achieved at 1,550 nm.

6. Consider a 100 nm thick indium tin oxide (ITO) film on a glass substrate. The refractive indices of the glass substrate and ITO are 1.48 and $1.815 - j0.0031$, respectively at $\lambda = 600$ nm. Find the reflection from this structure. Determine what film can be deposited on top of the ITO to reduce this reflection to zero at this wavelength.

7. A thin film of silicon is deposited on a silver metal substrate. The refractive index of silicon at 532 nm is $4.15 - j0.05$, and the refractive index of silver is $0.13 - j3.15$ at 532 nm. Using the effective reflectance index contours, find the thickness of silicon that will produce the lowest reflection at 532 nm.

4 Transfer Matrix Method

4.1 TRANSFER MATRIX METHOD FOR NORMAL INCIDENCE

The transfer matrix method (TMM) can be derived by writing the field expressions and considering their continuity across each boundary as derived in Chapter 1. Alternatively, it can also be derived by considering the Fresnel reflections at each boundary, followed by a propagation through each layer. Although both approaches produce identical results, the former approach is taken here because it uses the wave equation more systematically rather than via a phenomenological treatment of the interfaces.

Consider a substrate with a refractive index of n_s; a number of layers with refractive indices n_{f1}, n_{f2}, etc.; and an outside medium (air) with refractive index n_a, as shown in Figure 4.1. The reflection and transmission coefficients at normal incidence can be derived by ensuring the continuity of the electric field amplitudes and their derivatives at all interfaces. Complex indices are assumed to have a negative imaginary part for absorption and a positive imaginary part for amplification (though the latter case only occurs in the case of optical amplification).

Considering the first interface at z_3 (which transitions from air to film #2), the field continuity equations can be written as

$$e^{-jk_0 n_a z_3} + re^{+jk_0 n_a z_3} = Ce^{-jk_0 n_{f2} z_3} + De^{+jk_0 n_{f2} z_3}. \qquad (4.1)$$

The field derivative can be written as

$$-jk_0 n_a e^{-jk_0 n_a z_3} + rjk_0 n_a e^{+jk_0 n_a z_3} = -jk_0 n_{f2} Ce^{-jk_0 n_{f2} z_3} + jk_0 n_{f2} De^{+jk_0 n_{f2} z_3}. \qquad (4.2)$$

The term jk_0 can be cancelled on both sides of equation (4.2), resulting in

$$n_a e^{-jk_0 n_a z_3} - rn_a e^{+jk_0 n_a z_3} = n_{f2} Ce^{-jk_0 n_{f2} z_3} - n_{f2} De^{+jk_0 n_{f2} z_3}. \qquad (4.3)$$

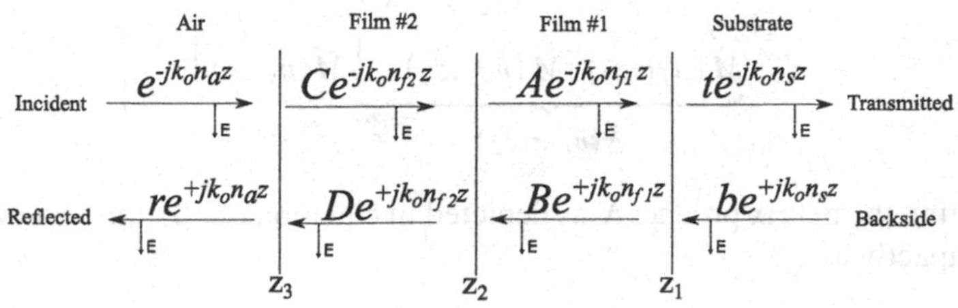

Figure 4.1 TMM with two thin films on a substrate.

Equations (4.1) and (4.3) can be combined into a 2×2 matrix form as

$$\begin{bmatrix} e^{-jk_0 n_a z_3} & e^{+jk_0 n_a z_3} \\ n_a e^{-jk_0 n_a z_3} & -n_a e^{+jk_0 n_a z_3} \end{bmatrix} \begin{bmatrix} 1 \\ r \end{bmatrix}$$
$$= \begin{bmatrix} e^{-jk_0 n_{f2} z_3} & e^{+jk_0 n_{f2} z_3} \\ n_{f2} e^{-jk_0 n_{f2} z_3} & -n_{f2} e^{+jk_0 n_{f2} z_3} \end{bmatrix} \begin{bmatrix} C \\ D \end{bmatrix}. \qquad (4.4)$$

This can be expressed more compactly as

$$M(n_a, z_3) \begin{bmatrix} 1 \\ r \end{bmatrix} = M(n_{f2}, z_3) \begin{bmatrix} C \\ D \end{bmatrix}. \qquad (4.5)$$

Next, moving to the interface at z_2 (film #2 to film #1), the relationship becomes

$$M(n_{f2}, z_2) \begin{bmatrix} C \\ D \end{bmatrix} = M(n_{f1}, z_2) \begin{bmatrix} A \\ B \end{bmatrix}, \qquad (4.6)$$

and the last interface at $z = z_1$ (film #2 to substrate) becomes

$$M(n_{f1}, z_1) \begin{bmatrix} A \\ B \end{bmatrix} = M(n_s, z_1) \begin{bmatrix} t \\ b \end{bmatrix}. \qquad (4.7)$$

We can solve for r and t by eliminating all the field amplitudes A, B, C, D, etc. except $\begin{bmatrix} 1 \\ r \end{bmatrix}$ and $\begin{bmatrix} t \\ b \end{bmatrix}$. Referring back to equation (4.5), we can also write it as

$$\begin{bmatrix} 1 \\ r \end{bmatrix} = [M(n_a, z_3)]^{-1} M(n_{f2}, z_3) \begin{bmatrix} C \\ D \end{bmatrix}. \qquad (4.8)$$

Then we can substitute equation (4.6) to get

$$\begin{bmatrix} 1 \\ r \end{bmatrix} = [M(n_a, z_3)]^{-1} M(n_{f2}, z_3) [M(n_{f2}, z_2)]^{-1} M(n_{f1}, z_2) \begin{bmatrix} A \\ B \end{bmatrix}. \qquad (4.9)$$

Substituting equation (4.7) results in

$$\begin{bmatrix} 1 \\ r \end{bmatrix} = [M(n_a, z_3)]^{-1} \underbrace{M(n_{f2}, z_3) [M(n_{f2}, z_2)]^{-1}}_{N(n_{f2}, z_2, z_3)}$$
$$\underbrace{M(n_{f1}, z_2) [M(n_{f1}, z_1)]^{-1}}_{N(n_{f1}, z_1, z_2)} M(n_s, z_1) \begin{bmatrix} t \\ b \end{bmatrix}. \qquad (4.10)$$

Representing the matrix product N as depicted in equation (4.10), this can be written more compactly as

$$\begin{bmatrix} 1 \\ r \end{bmatrix} = [M(n_a, z_3)]^{-1} N(n_{f2}, z_2, z_3) N(n_{f1}, z_1, z_2) M(n_s, z_1) \begin{bmatrix} t \\ b \end{bmatrix}. \qquad (4.11)$$

Since all of these are 2×2 matrices, the product of all these will also be a 2×2 matrix. Therefore, we can write the whole expression as

$$\begin{bmatrix} 1 \\ r \end{bmatrix} = M \begin{bmatrix} t \\ b \end{bmatrix}, \tag{4.12}$$

where M is the 2×2 product of all the matrices shown in equation (4.10) and r, t, and b are the reflection, transmission, and backside-incident coefficients, respectively. For now, we can assume that b is zero because usually there will be no field incident from the backside, unless there is an external feedback that needs to be considered separately (more on that later). Therefore,

$$\begin{bmatrix} 1 \\ r \end{bmatrix} = M \begin{bmatrix} t \\ 0 \end{bmatrix}. \tag{4.13}$$

From equation (4.13), we can get

$$t = \frac{1}{M_{11}} \tag{4.14}$$

and

$$r = M_{21} t \tag{4.15}$$

$$= \frac{M_{21}}{M_{11}}. \tag{4.16}$$

The reflection and transmission of the power flux, R and T, can be computed by considering the ratio between the incident and reflected (or transmitted) power fluxes

$$R = \frac{S_r}{S_i} \tag{4.17}$$

$$T = \frac{S_t}{S_i}, \tag{4.18}$$

where S_i, S_r, and S_t are the power fluxes. Using the expressions derived earlier for the directional power flux of a plane wave (equation (1.68) in Chapter 1), we can get

$$S_i = \frac{|A_i|^2}{2} \sqrt{\frac{\varepsilon_0}{\mu_0}} n_a \tag{4.19}$$

$$S_r = \frac{|A_r|^2}{2} \sqrt{\frac{\varepsilon_0}{\mu_0}} n_a \tag{4.20}$$

$$S_t = \frac{|A_t|^2}{2} \sqrt{\frac{\varepsilon_0}{\mu_0}} n_s, \tag{4.21}$$

where n_a and n_s are the refractive indices of the incident medium (air) and transmitted medium (substrate), respectively. Therefore, we can finally get

$$R = \frac{S_r}{S_i} = \frac{|A_r|^2}{|A_i|^2} = \left| \frac{A_r}{A_i} \right|^2 = |r|^2 = \left| \frac{M_{21}}{M_{11}} \right|^2 \tag{4.22}$$

and

$$T = \frac{S_t}{S_i} = \frac{n_s}{n_a}\left|\frac{A_t}{A_i}\right|^2 = \frac{n_s}{n_a}|t|^2 = \frac{n_s}{n_a}\left|\frac{1}{M_{11}}\right|^2. \tag{4.23}$$

Equations (4.22) and (4.23) are the basic outcomes of the plane-wave TMM. They allow us to compute the reflection and transmission as a function of wavelength for any number of layers. The method also allows for the refractive indices of any layer to be complex without any additional modifications. However, because we assumed the incident medium (air) and the transmitted medium (substrate) to have infinite thickness, their indices must remain real. If we are not interested in calculating transmission, then the substrate can be complex.

The total absorption A in the structure can also be calculated using

$$A = 1 - R - T. \tag{4.24}$$

4.2 INCLUDING THE EFFECTS OF REFLECTION FROM THE BACKSIDE OF THE SUBSTRATE

By setting $b = 0$ in equation (4.13), we have implicitly ignored any reflection from the backside of the substrate. In practice, there will be reflection from the substrate/air interface on the backside of the substrate. It might be tempting to treat the substrate as another film and treat the air behind it as the substrate. But this will not work. Because the substrate is usually much thicker than the coherence length of ordinary light, the forward and backward moving fields will not interfere like they do inside a thin film. Instead, the forward and backward fields must be combined incoherently [1,2].

Referring to Figure 4.2, we will assume that the backside of the substrate has a reflection value of R_s and a transmission value of T_s. We can trace the values of reflected power from each bounce off the backside of the substrate and add them incoherently. If the incident power flux has a magnitude of 1, the transmitted power flux into the substrate will be T (as computed from equation (4.23)). This will be reflected by the substrate to produce a power of R_sT incident from the right hand side (substrate side) of the film stack. If we can calculate the reflection R_r and transmission T_r of the film stack in the reverse direction (incident from the substrate and exiting through the front side), we can get the transmitted power on the front side due to the first bounce from the substrate to be $T_rR_sTe^{-4k_o\kappa_sL}$, where the term $e^{-2k_o\kappa_sL}$ is the absorption loss in the substrate due to a single pass and κ_s is the imaginary part of the substrate's refractive index. Similarly, the second bounce will contribute $T_rR_sR_rR_sTe^{-8k_o\kappa_sL}$. We can see that the total reflection R_t on the incident side can be expressed as a geometric series

$$R_t = R + T_rR_sTe^{-4k_o\kappa_sL} + T_rR_sR_rR_sTe^{-8k_o\kappa_sL} \ldots \tag{4.25}$$

$$= R + T_rR_sTe^{-4k_o\kappa_sL}\left(1 + R_rR_se^{-4k_on_{si}L}\ldots\right) \tag{4.26}$$

$$= R + T_rR_sTe^{-4k_o\kappa_sL}\sum_{n=0}^{\infty}\left(R_rR_se^{-4k_o\kappa_sL}\right)^n \tag{4.27}$$

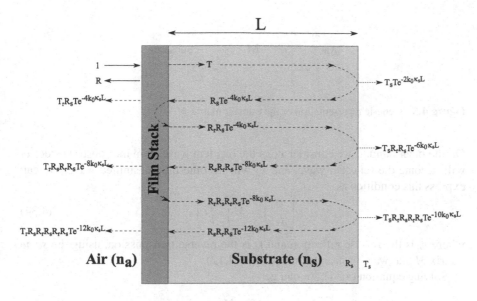

Figure 4.2 TMM including the reflection from the backside of the substrate.

$$= R + \frac{T_r R_s T e^{-4k_o \kappa_s L}}{1 - R_r R_s e^{-4k_o \kappa_s L}}. \tag{4.28}$$

The transmission through the backside of the substrate can be similarly derived as

$$T_t = T_s T e^{-2k_o \kappa_s L} + T_s R_r R_s T e^{-6k_o \kappa_s L} \dots \tag{4.29}$$

$$= T_s T e^{-2k_o \kappa_s L} \left(1 + R_r R_s e^{-4k_o \kappa_s L} \dots \right) \tag{4.30}$$

$$= T_s T e^{-2k_o \kappa_s L} \sum_{n=0}^{\infty} \left(R_r R_s e^{-4k_o \kappa_s L} \right)^n \tag{4.31}$$

$$= \frac{T_s T e^{-2k_o \kappa_s L}}{1 - R_r R_s e^{-4k_o \kappa_s L}}. \tag{4.32}$$

In order to evaluate these expressions, we need to know the reverse reflection R_r and reverse transmission T_r due to a field entering the film stack from the substrate side and exiting through the front side. To get an expression for this, we can write equation (4.12) more generically as

$$\begin{bmatrix} x_1 \\ x_2 \end{bmatrix} = M \begin{bmatrix} x_3 \\ x_4 \end{bmatrix}. \tag{4.33}$$

As illustrated in Figure 4.3, if x_1 is the incident wave from the air side, x_2 will be the reflected wave, and x_3 will be the transmitted wave. This definition resulted in equation (4.13), which is

$$\begin{bmatrix} 1 \\ r \end{bmatrix} = M \begin{bmatrix} t \\ 0 \end{bmatrix}.$$

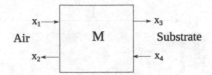

Figure 4.3   Generic representation of the transfer matrix M.

On the other hand, if we consider x_4 as the incident wave from the substrate side, x_3 will become the reflected wave, and x_2 will become the transmitted wave. We can express this condition as

$$\begin{bmatrix} 0 \\ t_r \end{bmatrix} = M \begin{bmatrix} r_r \\ 1 \end{bmatrix}, \tag{4.34}$$

where r_r is the reverse reflection and t_r is the reverse transmission, using the *same* matrix M that was constructed in equation (4.12).

Solving equation (4.34), we can get

$$r_r = -\frac{M_{12}}{M_{11}} \tag{4.35}$$

and

$$t_r = M_{22} - \frac{M_{21}M_{12}}{M_{11}}. \tag{4.36}$$

Therefore,

$$R_r = \left| \frac{M_{12}}{M_{11}} \right|^2 \tag{4.37}$$

and

$$T_r = \frac{n_a}{n_s} \left| M_{22} - \frac{M_{21}M_{12}}{M_{11}} \right|^2. \tag{4.38}$$

Now that we know the expressions for R_r and T_r, we can easily evaluate the total reflection R_t and total transmission T_t from equations (4.28) and (4.32), which include the incoherent contribution from the backside of the substrate.

4.3   EXAMPLE – ANTIREFLECTION ON SILICA GLASS

Consider a silica substrate with a refractive index of 1.48 coated on one side with a single-layer antireflection film with an index of 1.217 at a reference wavelength of 550 nm. Considering the wavelength range of 400–800 nm with 1,000 interval points, we can evaluate R and T using equations (4.22) and (4.23) for each wavelength point. This ignores the effect of the substrate reflection. We can also compute R_t from equation (4.28) which includes the effects of the incoherent substrate reflection. Both of these are shown in Figure 4.4. The computer code for producing this plot is given in Chapter 15.

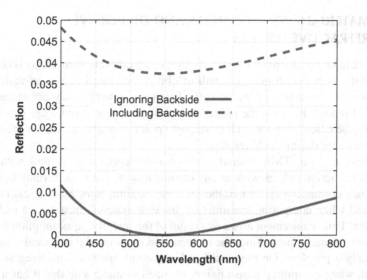

Figure 4.4 Example TMM calculation of a single-layer antireflection coating on a silica glass substrate with and without the consideration of backside reflection.

As expected, the reflection falls to zero at the reference wavelength of 550 nm. As we can see, the backside reflection offsets the entire reflection spectrum by about 4%. This can be reduced by using an identical antireflection layer on the backside of the substrate.

4.4   FILM STACKS ON BOTH SIDES OF THE SUBSTRATE

When both the front side and backside of the substrate are coated with thin films (which may be of the same design or different designs) using the formulation developed in Section 4.2, it is relatively straightforward to calculate the overall performance. Instead of R_s representing just the reflection from the substrate/air interface, it should now capture the effects of all the films on the backside. This R_s has to be calculated for an incident beam entering from the substrate side, not from the air side. This can be easily calculated by modeling the backside stack exactly like the front side stack but using equation (4.37) to get the reflection from the substrate side. That is,

$$R_s = \left| \frac{M_{12}^{B}}{M_{11}^{B}} \right|^2 , \qquad (4.39)$$

where M^B is the transfer matrix for the backside film stack.

The TMM code given in Chapter 15 does not include computation of backside stack (but it does include backside substrate/air interface). This is done to keep the example simple and easier to follow. The reader is encouraged to examine ways of including these additional features into their code.

4.5   MATERIALS WITH COMPLEX AND DISPERSIVE REFRACTIVE INDICES

So far, we have considered the refractive index as a constant value. This is obviously not realistic because all materials will exhibit dispersion; i.e., the refractive index value will be a function of wavelength. This can be very easily accounted for in the TMM model. Because the reflection and transmission spectra are produced by repeating the calculation for each wavelength point, it is a trivial matter to adjust the refractive index during each iteration.

Additionally, our TMM formulation is equally applicable to films with real or complex refractive indices without any modification to the basic formulation. However, when computing reflection, the incident medium must have a real refractive index, and when computing transmission, the exit material must have a real refractive index. This requirement arises as a result of the underlying assumption that these media are infinitely thick. Since the incident and exit material is typically air, this is not usually a problem. On the other hand, if the substrate is considered as the exit material, when computing transmission, we need to make sure that it has a real refractive index. Alternatively, it can also be treated as a very thick incoherent film as discussed in Section 4.2, in which case it is allowed to have a complex refractive index. Similarly, the incident medium can also be a thick incoherent layer (such as, for example, a layer of water). A formulation similar to the one described in Section 4.2 can be developed for this scenario as well.

As an example, Figure 4.5 shows the reflection spectrum from three layers of SiO_2, Si, and Cu on a Si substrate as indicated in the figure inset. This is essentially what is known as the SOI (silicon on insulator) substrate with a thin copper

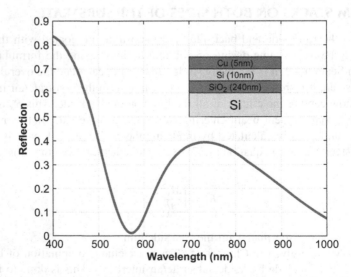

Figure 4.5   Demonstration of TMM calculation using three layers of dispersive complex materials on a silicon substrate.

film on top. The purpose of this plot is not to illustrate a design principle but to show how TMM can be used with dispersive and complex films. The code used for producing this plot is shown in Chapter 15.

4.6 CALCULATION OF ABSORPTION IN FILMS

The net absorption in a multilayer structure can be calculated by evaluating equation (4.24). This will produce the *total* absorption from all layers. Although this is a very straightforward calculation, there are instances when we need to calculate the optical absorption inside specific layers. This can be done by calculating the difference between the directional power flux entering and exiting the film. This is represented by the divergence of **S**, which was defined earlier as

$$\nabla \cdot \mathbf{S} = -P. \tag{4.40}$$

Referring to Figure 4.1, and using the definition of S from equation (1.68) in Chapter 1, we can calculate the power being absorbed in, for example, film #2 as

$$P_{\text{film #2}} = |S_{z_3}| - |S_{z_2}| \tag{4.41}$$

$$= \frac{1}{2}\sqrt{\frac{\varepsilon_0}{\mu_0}} |n_{f2}| \left[\left(1 - e^{-2k_0\kappa(z_3-z_2)}\right) |C|^2 - \left(e^{-2k_0\kappa(z_3-z_2)} - 1\right) |D|^2 \right] \tag{4.42}$$

$$= \frac{1}{2}\sqrt{\frac{\varepsilon_0}{\mu_0}} |n_{f2}| \left[\left(1 - e^{-2k_0\kappa(z_3-z_2)}\right) \left(|C|^2 + |D|^2\right) \right], \tag{4.43}$$

where κ is the imaginary part of the refractive index of film #2. When normalized against the incident power, this expression becomes

$$\frac{P_{\text{film #2}}}{P_{\text{inc}}} = \frac{\frac{1}{2}\sqrt{\frac{\varepsilon_0}{\mu_0}} |n_{f2}| \left[\left(1 - e^{-2k_0\kappa(z_3-z_2)}\right) \left(|C|^2 + |D|^2\right) \right]}{\frac{1}{2}\sqrt{\frac{\varepsilon_0}{\mu_0}} |n_a|} \tag{4.44}$$

$$= \frac{|n_{f2}|}{|n_a|} \left(1 - e^{-2k_0\kappa(z_3-z_2)}\right) \left(|C|^2 + |D|^2\right). \tag{4.45}$$

Therefore, we can calculate the percentage of power absorbed in any film by considering the field amplitude coefficients of the forward and backward waves, along with the complex refractive index of the film.

4.7 CALCULATION OF THE FIELD DISTRIBUTION

The spatial field distribution inside the film stack can be calculated using the same TMM. This is useful for understanding the optical confinement and resonances taking place in the film stack.

The field distribution in each layer was earlier assumed to be

$$\text{Air}: \quad F_a = e^{-jk_0 n_a z} + r e^{+jk_0 n_a z} \tag{4.46}$$

$$\text{Film } \#2: \quad F_2 = Ce^{-jk_0 n_{f2} z} + De^{+jk_0 n_{f2} z} \tag{4.47}$$

$$\text{Film } \#1: \quad F_1 = Ae^{-jk_0 n_{f1} z} + Be^{+jk_0 n_{f1} z} \tag{4.48}$$

$$\text{Substrate}: \quad F_s = te^{-jk_0 n_s z} + be^{+jk_0 n_s z}. \tag{4.49}$$

In order to compute the field distribution, we have to first calculate the field reflection coefficient r at the desired wavelength. Once r is known, we can use equation (4.46) to calculate the field distribution in the incident medium (air). Then, r can be used in equation (4.5) to calculate the field amplitude coefficients C and D in film #2, resulting in

$$\begin{bmatrix} C \\ D \end{bmatrix} = [M(n_{f2}, z_3)]^{-1} M(n_a, z_3) \begin{bmatrix} 1 \\ r \end{bmatrix}. \tag{4.50}$$

This allows us to calculate the field distribution F_2 in equation (4.47).

Using C and D, we can then calculate the field amplitudes A and B in film #1. From equation (4.6), this becomes

$$\begin{bmatrix} A \\ B \end{bmatrix} = [M(n_{f1}, z_2)]^{-1} M(n_{f2}, z_2) \begin{bmatrix} C \\ D \end{bmatrix}. \tag{4.51}$$

As a result, the field distribution F_1 can be computed.

Finally, in the substrate, we can get

$$\begin{bmatrix} t \\ b \end{bmatrix} = [M(n_s, z_1)]^{-1} M(n_{f1}, z_1) \begin{bmatrix} A \\ B \end{bmatrix}. \tag{4.52}$$

If r was originally computed with no backside reflection, then b from the above equation should automatically work out to be zero.

4.7.1 EXAMPLE – FIELD DISTRIBUTION IN THE SINGLE-LAYER ANTIREFLECTION STRUCTURE

Consider the antireflection example studied in Section 4.3. The field distribution at the antireflection wavelength of $\lambda = 550$ nm is shown in Figure 4.6. We can notice several important things about this plot. The field profile in the incident medium (air) is flat with an amplitude of 1.0. This is consistent with the antireflection condition because the incident field will have a magnitude of 1.0 and the reflected field will have a magnitude of zero. Since there is only one field component, there will be no interference, and the profile will be uniform. The substrate also has a flat field profile. Again, this is due to the presence of only one field component – because b is zero, we will only have the forward moving (transmitted) field component. Notice that the transmitted field amplitude is less than 1.0, even though the incident and transmitted power fluxes are equal. The reason for this should be evident from equation (4.23). Because the refractive index of the substrate is higher than air, the field amplitude in the substrate will be smaller to preserve the power flux. The films, on the other hand, will contain both the forward and backward moving field components. As a result, there will be interference between these components. This is exhibited by the smoothly varying amplitude inside the film as shown in Figure 4.6.

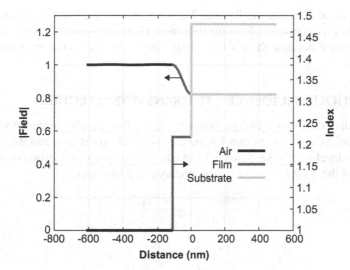

Figure 4.6 Field distribution in the antireflection structure discussed in Figure 4.4 at the reference wavelength of $\lambda = 550$ nm.

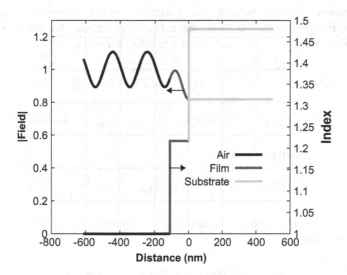

Figure 4.7 Field distribution in the antireflection structure discussed in Figure 4.4 at $\lambda = 400$ nm.

Instead of plotting the field at the reference wavelength of 550 nm, if we plot the field profile at $\lambda = 400$ nm, the distribution will be as shown in Figure 4.7. The most important difference we can see here is the large field oscillations in the incident medium (air). This is due to the nonzero reflection at this wavelength. Due to this reflection, both the forward and backward field components will contribute to interference oscillations. The magnitude of the oscillations, obviously, will depend on the magnitude of the reflection. However, the field profile is still flat in the substrate because there will be only one component – the transmitted field component.

4.8 OBLIQUE INCIDENCE – TE (TRANSVERSE ELECTRIC)

If the incident radiation is oblique with the electric field parallel to the interface (TE polarization) as shown in Figure 4.8, we can use field expressions and the boundary conditions developed in Section 1.5.3.1 of Chapter 1 to derive the expressions for the elements of the TMM matrices. The boundary conditions were

$$\left.\frac{\partial E_y}{\partial x}\right|_{z=0^-} = \left.\frac{\partial E_y}{\partial x}\right|_{z=0^+} \tag{4.53}$$

$$\left.\frac{\partial E_y}{\partial z}\right|_{z=0^-} = \left.\frac{\partial E_y}{\partial z}\right|_{z=0^+} \tag{4.54}$$

and

$$\left. E_y \right|_{z=0^-} = \left. E_y \right|_{z=0^+}. \tag{4.55}$$

The expressions for the incident and transmitted plane waves were derived in Chapter 1. Following this method, the expressions inside the matrices can be obtained. For example, at $z = z_3$, the matrix relation becomes

$$
\begin{bmatrix}
e^{-jk_0\sqrt{n_a^2-n_a^2\sin^2\theta_a}\,z_3} & e^{+jk_0\sqrt{n_a^2-n_a^2\sin^2\theta_a}\,z_3} \\
\sqrt{n_a^2-n_a^2\sin^2\theta_a}\,e^{-jk_0\sqrt{n_a^2-n_a^2\sin^2\theta_a}z_3} & -\sqrt{n_a^2-n_a^2\sin^2\theta_a}\,e^{+jk_0\sqrt{n_a^2-n_a^2\sin^2\theta_a}\,z_3}
\end{bmatrix}
\begin{bmatrix} 1 \\ r \end{bmatrix} =
$$
$$
\begin{bmatrix}
e^{-jk_0\sqrt{n_{f2}^2-n_a^2\sin^2\theta_a}\,z_3} & e^{-jk_0\sqrt{n_{f2}^2-n_a^2\sin^2\theta_a}\,z_3} \\
\sqrt{n_{f2}^2-n_a^2\sin^2\theta_a}\,e^{-jk_0\sqrt{n_{f2}^2-n_a^2\sin^2\theta_a}\,z_3} & -\sqrt{n_{f2}^2-n_a^2\sin^2\theta_a}\,e^{-jk_0\sqrt{n_{f2}^2-n_a^2\sin^2\theta_a}\,z_3}
\end{bmatrix}
\begin{bmatrix} C \\ D \end{bmatrix}, \tag{4.56}
$$

which can be compactly written as

$$M^{\mathrm{TE}}(n_a, z_3) \begin{bmatrix} 1 \\ r \end{bmatrix} = M^{\mathrm{TE}}(n_{f2}, z_3) \begin{bmatrix} C \\ D \end{bmatrix}. \tag{4.57}$$

Using this definition, we can reconstruct the overall transfer matrix M that we defined in equation (4.12).

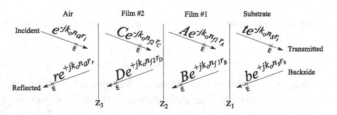

Figure 4.8 TMM for oblique incidence (TE). All electric fields are pointing into the plane of the paper.

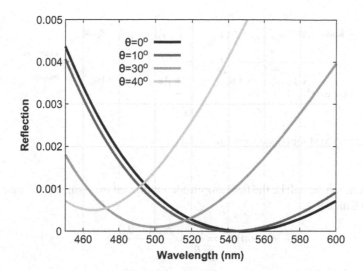

Figure 4.9 Quarter-wave antireflection design using $n_f = 1.217$ that was discussed in Figure 4.4 as a function of incidence angles for TE polarization.

Figure 4.9 shows the calculated reflection spectra of the single-layer antireflection design shown in Figure 4.4. At normal incidence, the reflection drops to zero at the reference wavelength of 550 nm as expected. When the incidence angle increases, the minimum reflection point moves to a shorter wavelength, and the antireflection becomes less effective. At 40° incidence, the minimum reflection is 0.05%, and the wavelength moves to 465 nm. This blue-shift with increasing angle of incidence occurs in all thin-film structures that were designed for normal incidence. It is also possible to design the structure so that its optimum performance will coincide with a specific angle of incidence. This aspect will be discussed in greater detail in Chapter 11.

4.9 OBLIQUE INCIDENCE – TM (TRANSVERSE MAGNETIC)

The continuity expressions of the fields and their derivatives for TM incidence were derived in Chapter 1. Unlike TE, in this case, the field had E_z and E_x components, as shown in Figure 4.10. We can write the transfer matrix relationship in terms of E_z or in terms of E_x. Both of them result in identical numerical results, although the expressions are different.

From Chapter 1, the E_z field component would satisfy the following boundary conditions:

$$\frac{\partial E_z}{\partial z}\bigg|_{z=0^-} = \frac{\partial E_z}{\partial z}\bigg|_{z=0^+} \tag{4.58}$$

$$E_z\left(0^+\right)\varepsilon_{f2} = E_z\left(0^-\right)\varepsilon_{f1}. \tag{4.59}$$

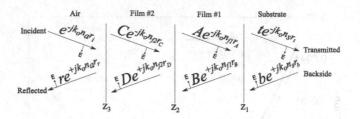

Figure 4.10 TMM for oblique incidence (TM).

The E_z component will be the field amplitude multiplied by $-\sin(\theta)$. Using equation (1.93) in Chapter 1,

$$\sin(\theta) = \frac{n_x}{n} = \frac{n_a \sin(\theta_a)}{n_f}. \tag{4.60}$$

The resulting matrix expression for the $z = z_3$ interface would be

$$\begin{bmatrix} n_a^2 \sin(\theta_a)\, e^{-jk_0\sqrt{n_a^2 - n_a^2 \sin^2\theta_a}\, z_3} & n_a^2 \sin(\theta_a)\, e^{+jk_0\sqrt{n_a^2 - n_a^2 \sin^2\theta_a}\, z_3} \\ \frac{n_a \sin(\theta_a)}{n_a}\sqrt{n_a^2 - n_a^2 \sin^2\theta_a}\, e^{-jk_0\sqrt{n_a^2 - n_a^2 \sin^2\theta_a}\, z_3} & -\frac{n_a \sin(\theta_a)}{n_a}\sqrt{n_a^2 - n_a^2 \sin^2\theta_a}\, e^{+jk_0\sqrt{n_a^2 - n_a^2 \sin^2\theta_a}\, z_3} \end{bmatrix} \begin{bmatrix} 1 \\ r \end{bmatrix} =$$

$$\begin{bmatrix} n_a n_{f2} \sin(\theta_a)\, e^{-jk_0\sqrt{n_{f2}^2 - n_a^2 \sin^2\theta_a}\, z_3} & n_a n_{f2} \sin(\theta_a)\, e^{+jk_0\sqrt{n_{f2}^2 - n_a^2 \sin^2\theta_a}\, z_3} \\ \frac{n_a \sin(\theta_a)}{n_{f2}}\sqrt{n_{f2}^2 - n_a^2 \sin^2\theta_a}\, e^{-jk_0\sqrt{n_{f2}^2 - n_a^2 \sin^2\theta_a}\, z_3} & -\frac{n_a \sin(\theta_a)}{n_{f2}}\sqrt{n_{f2}^2 - n_a^2 \sin^2\theta_a}\, e^{-jk_0\sqrt{n_{f2}^2 - n_a^2 \sin^2\theta_a}\, z_3} \end{bmatrix} \begin{bmatrix} C \\ D \end{bmatrix}. \tag{4.61}$$

One pitfall with the above boundary condition is that the matrices will fail under normal incidence. Since the E_z field component will become zero when the incidence angle is zero, all of the matrix elements will collapse to zero, resulting in a singularity.

An alternative approach is to express the TM boundary condition in terms of E_x. Since E_x becomes equal to E under normal incidence, it can be used for all angles including normal incidence. From Chapter 1, the E_x field component satisfied the following two equations:

$$E_x|_{z=0^-} = E_x|_{z=0^+} \tag{4.62}$$

and

$$\left.\frac{\partial E_x}{\partial z}\right|_{z=0^-} - \tan(\theta_i)\frac{\partial E_x}{\partial x} = \left.\frac{\partial E_x}{\partial z}\right|_{z=0^+} - \tan(\theta_t)\frac{\partial E_x}{\partial x}. \tag{4.63}$$

The E_x component is the field amplitude multiplied by $\cos(\theta)$. Using the expression derived for $\sin(\theta)$ in equation (1.93) of Chapter 1, we can write $\tan(\theta)$ in each film as

$$\tan(\theta) = \frac{n_x}{\sqrt{n_f^2 - n_x^2}} = \frac{n_a \sin\theta_a}{\sqrt{n_f^2 - n_a^2 \sin^2(\theta_a)}}. \tag{4.64}$$

Using equations (4.62) and (4.63) in the transfer matrix, we can rewrite all of the matrices. For instance, at the $z = z_3$ interface,

$$
\begin{bmatrix} \dfrac{\sqrt{n_a{}^2-n_a{}^2\sin^2\theta_a}}{n_a}\, e^{-jk_0\sqrt{n_a{}^2-n_a{}^2\sin^2\theta_a}\,z_3} & -\dfrac{\sqrt{n_a{}^2-n_a{}^2\sin^2\theta_a}}{n_a}\, e^{+jk_0\sqrt{n_a{}^2-n_a{}^2\sin^2\theta_a}\,z_3} \\[2mm] n_a e^{-jk_0\sqrt{n_a{}^2-n_a{}^2\sin^2\theta_a}\,z_3} & n_a e^{+jk_0\sqrt{n_a{}^2-n_a{}^2\sin^2\theta_a}\,z_3} \end{bmatrix} \begin{bmatrix} 1 \\ r \end{bmatrix} =
$$
$$
\begin{bmatrix} \dfrac{\sqrt{n_{f2}{}^2-n_a{}^2\sin^2\theta_a}}{n_{f2}}\, e^{-jk_0\sqrt{n_{f2}{}^2-n_a{}^2\sin^2\theta_a}\,z_3} & -\dfrac{\sqrt{n_{f2}{}^2-n_a{}^2\sin^2\theta_a}}{n_{f2}}\, e^{+jk_0\sqrt{n_{f2}{}^2-n_a{}^2\sin^2\theta_a}\,z_3} \\[2mm] n_{f2} e^{-jk_0\sqrt{n_{f2}{}^2-n_a{}^2\sin^2\theta_a}\,z_3} & n_{f2} e^{-jk_0\sqrt{n_{f2}{}^2-n_a{}^2\sin^2\theta_a}\,z_3} \end{bmatrix} \begin{bmatrix} C \\ D \end{bmatrix}.
$$
$$(4.65)$$

Both equations (4.61) and (4.65) produce identical results. Whichever we choose to use, this system can be written as

$$
M^{\text{TM}}(n_a, z_3) \begin{bmatrix} 1 \\ r \end{bmatrix} = M^{\text{TM}}(n_{f2}, z_3) \begin{bmatrix} C \\ D \end{bmatrix}. \tag{4.66}
$$

Similar to the TE case, we can reconstruct the overall transfer matrix M in equation (4.12) to solve the overall system response.

Figure 4.11 shows the computed results for different angles of incidence using this formulation for TM polarization. The behavior is similar to the TE polarization, with a blue shift and a decrease in the antireflection performance. There are some important differences between TE and TM, and one of them is when the light is incident at Brewster's angle. At this angle, the reflection falls to zero for TM, but not for TE, which can be exploited to make polarization-selective filters. This aspect will be explored in greater detail in Chapter 11.

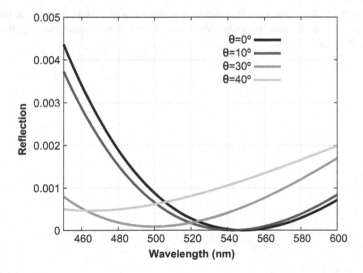

Figure 4.11   Quarter-wave antireflection design using $n_f = 1.217$ as discussed in Figure 4.4 as a function of incidence angles for TM polarization.

4.10 PROBLEMS

1. Assuming a 100 nm indium tin oxide (ITO) film (index of $1.89 - j0.0023$ at a wavelength of 600 nm) is on a SiO_2 substrate, calculate the reflection, transmission, and absorption using TMM.
2. A 200 nm TiO_2 film is on a 1 mm thick fused silica glass substrate. Using the complex refractive index dispersion of the materials, calculate the reflection, transmission, and absorption spectra between 350 and 500 nm wavelengths, including the backside reflections from the substrate.
3. A 200 nm SiO film is deposited on a Si substrate. Plot the field profiles for this structure at $\lambda = 1,500$ nm, and verify that it operates as an antireflection film.
4. An SOI wafer consists of a silicon substrate, 350 nm of SiO_2, and 150 nm of Si, in that order. If $\lambda = 532$ nm light is incident on this SOI wafer, calculate the fraction of optical absorption taking place in the top silicon film of the SOI wafer.
5. Modify the TMM code to include a film stack on the backside of the substrate using the formulation suggested in Section 4.4. For a silicon substrate with a single-layer antireflection coating applied on both sides of the substrate at a reference wavelength of 1,550 nm, verify that the result produces zero reflection.

REFERENCES

1. Larouche, S. & Martinu, L. OpenFilters: open-source software for the design, optimization, and synthesis of optical filters. *Applied Optics* **47**, C219. ISBN: 0003-6935 (May 2008).
2. Katsidis, C. C. & Siapkas, D. I. General transfer-matrix method for optical multilayer systems with coherent, partially coherent, and incoherent interference. *Applied Optics* **41**, 3978 (July 2002).

5 Multilayer Antireflection Theory

5.1 TWO-LAYER QUARTER-WAVE ANTIREFLECTION DESIGNS

The reflection from multiple quarter-wave films can be calculated by simply extending the single-film reflection calculations described in Chapter 3. Using one film with a phase thickness of $\theta = \frac{\pi}{2}$, we found that the effective reflectance index of the substrate became $n_r = \frac{n_f^2}{n_s}$. When dealing with more than one film, we will assign a subscript to keep track of each film. The effect of adding a film with $\theta_1 = \frac{\pi}{2}$ on the substrate will be $n_{r1} = \frac{n_{f1}^2}{n_s}$, where n_{f1} is the index of the film and n_{r1} is the resulting effective reflectance index. When a second film of index n_{f2} and a phase of $\theta_2 = \frac{\pi}{2}$ is deposited on top of the first film, the effective reflectance index will become $n_{r2} = \frac{n_{f2}^2}{n_{r1}} = \frac{n_{f2}^2}{n_{f1}^2}n_s$. Using this, the overall reflection coefficient with both films on the substrate can be written as

$$r = \frac{n_{r2} - n_a}{n_{r2} + n_a} = \frac{\frac{n_{f2}^2}{n_{f1}^2}n_s - n_a}{\frac{n_{f2}^2}{n_{f1}^2}n_s + n_a}. \tag{5.1}$$

Now the required condition to eliminate reflection at the reference wavelength is

$$\frac{n_{f2}^2}{n_{f1}^2}n_s = n_a \tag{5.2}$$

which can also be written as

$$\frac{n_{f2}^2}{n_{f1}^2} = \frac{n_a}{n_s}. \tag{5.3}$$

Therefore, the condition that needs to be met is the *ratio* between the two film indices. This relaxes the requirement compared to the single-layer antireflection design and allows us to expand the choice of materials. We can now select any two materials as long as they can satisfy $\frac{n_{f2}}{n_{f1}} = \sqrt{\frac{n_a}{n_s}}$.

For a substrate with a refractive index of 1.48, from equation (5.3), we can infer that film #2 must have a lower index than film #1. For example, if we use MgF_2 with an index of 1.38 as film #2, the film #1 must have an index of 1.68. If n_r is plotted in a complex plane as before, it will describe two connected arcs as shown in Figure 5.1. The first arc starts at $n_s = 1.48$ and ends at $n_r = 1.91$ at $\theta_1 = \frac{\pi}{2}$. The second arc starts

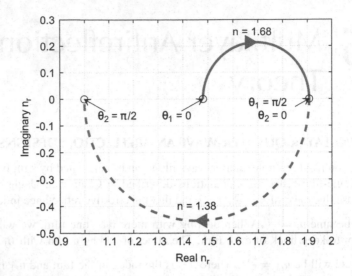

Figure 5.1 Effective reflective index contour of two quarter-wave films with $n_{f1} = 1.68$ and $n_{f2} = 1.38$ on a silica substrate.

at $n_r = 1.91$ and ends at $n_r = 1.0$ when $\theta_2 = \frac{\pi}{2}$. Since the final n_r is equal to 1.0, it satisfies the required condition for antireflection.

For film #1, one could consider Al_2O_3, which has an index of 1.77 at $\lambda = 550$ nm. Since its index is higher than the ideal value of 1.68, the effective reflectance index will be $\frac{1.38^2}{1.77^2}1.48 = 0.90$ instead of 1.0. Therefore, it won't be a perfect antireflection film. Nevertheless, we can estimate the reflection as $R = \left|\frac{1-0.90}{1+0.90}\right|^2 = 0.28\%$, which is still a reasonably small value. The calculated spectral reflection is shown in Figure 5.2, along with the ideal case of $n_{f1} = 1.68$.

On the other hand, there may be other reasons to use specific film materials. For instance, if film #1 has to be electrically conductive, we could consider ITO (indium tin oxide) instead of Al_2O_3. The refractive index of ITO at $\lambda = 550$ nm is about 1.85. This will result in an effective reflectance index of $\frac{1.38^2}{1.85^2}1.48 = 0.82$, and the reflection will be $R = \left|\frac{1-0.82}{1+0.82}\right|^2 = 0.98\%$.

We can also expand equation (5.3) to include more than two layers. The effective reflectance index of N quarter-wave layers becomes

$$n_{r,N} = \frac{n_{f,N}^2}{n_{f,N-1}^2}\frac{n_{f,N-2}^2}{n_{f,N-3}}\ldots n_s \quad \text{for even } N \tag{5.4}$$

and

$$n_{r,N} = \frac{n_{f,N}^2}{n_{f,N-1}^2}\frac{n_{f,N-2}^2}{n_{f,N-3}}\ldots\frac{1}{n_s} \quad \text{for odd } N. \tag{5.5}$$

Therefore, it is possible to come up with more elaborate layer structures and still satisfy the antireflection condition.

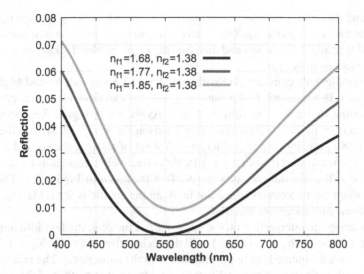

Figure 5.2 Spectral reflectance of the two-layer design with the exact solution of 1.68, Al_2O_3 (index of 1.77) and ITO (index of 1.85).

At this point, it is useful to introduce a simplified notation to represent the layer structure of our designs. In the above example, the notation that is typically used is silica$|HL|$air where H is the quarter-wave-thick high-index material and L is the quarter-wave-thick low-index material. This notation will become more useful as we expand our design space to include a large number of layers.

5.2 TWO-LAYER NON-QUARTER-WAVE ANTIREFLECTION DESIGNS

In the previous example, we used ITO and MgF_2 to reduce the reflection from 3.7% to 0.98%. Even though this is less than one-third of the uncoated substrate reflection, we can do better than this. It is possible to use these same two materials to achieve a perfect antireflection condition. This is done by relaxing the quarter-wave requirement (but still insisting on the final n_r being equal to 1.0).

For a single layer, we showed that the substrate index became modified to

$$n_{r1} = n_{f1} \frac{(n_s + n_{f1}) + (n_s - n_{f1}) e^{-2jk_0 n_{f1} z_1}}{(n_s + n_{f1}) - (n_s - n_{f1}) e^{-2jk_0 n_{f1} z_1}}. \tag{5.6}$$

After the second film, the effective reflectance index becomes

$$n_{r2} = n_{f2} \frac{(n_{r1} + n_{f2}) + (n_{r1} - n_{f2}) e^{-2jk_0 n_{f2} z_2}}{(n_{r1} + n_{f2}) - (n_{r1} - n_{f2}) e^{-2jk_0 n_{f2} z_2}}. \tag{5.7}$$

In order to achieve antireflection, only n_{r2} has to be real and equal to n_a. The value of n_{r1} does not have to satisfy any condition, nor does it have to be real. In other words, film #1 and film #2 do not have to be quarter wave thick, as long as their net effect makes the final n_{r2} equal to n_a. This is a very useful result because now we can start with readily available materials and then find the required film thicknesses to achieve the antireflection condition.

For example, let's consider ITO with an index of 1.85 as film #1 and MgF_2 with an index of 1.38 as film #2. In the complex plane, we can draw two complete circles corresponding to a half-wave film thickness. As shown in Figure 5.3, the first arc starts from $n_r = n_s = 1.48$, crosses the real axis again at $n_r = 2.31$, and then goes back to 1.48 at a half-wave film thickness. Instead of starting the second arc from $n_{r1} = 2.31$, we can start it from $n_{r2} = 1$ (which effectively treats air as its substrate). This circle will cross the real axis at 1.9. This is shown in Figure 5.3. The intercepts between the two circles (there are two) are the solutions for achieving perfect antireflection with these two materials.

Considering the intersection on the upper half of the plot, the first film must have a phase thickness of $\theta_{1a} = 0.418 \left(\frac{\pi}{2} \right)$, and the second film has to be $\theta_{2a} = 1.23 \left(\frac{\pi}{2} \right)$, where we have designated the upper solution with the subscript a. The resulting plot on the complex plane is shown in Figure 5.4a. Notice that neither of the arcs are a quarter wave thick. This technique allows us to use existing materials to meet the perfect antireflection condition. An equally valid solution could have been obtained by selecting the intersection on the lower half of Figure 5.3. Representing this solution with b, we can get $\theta_{1b} = 1.582 \left(\frac{\pi}{2} \right)$ for the first film and $\theta_{2b} = 0.7689 \left(\frac{\pi}{2} \right)$ for the second film. This is shown Figure 5.4b.

The spectral reflection for both designs computed using the transfer matrix method (TMM) is shown in Figure 5.5. As expected, their performance at the reference wavelength $\lambda_0 = 550$ nm is the same in both cases.

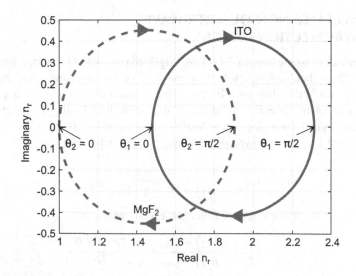

Figure 5.3 Contour plots of half-wave ITO on silica and half-wave MgF_2 on air as the substrate.

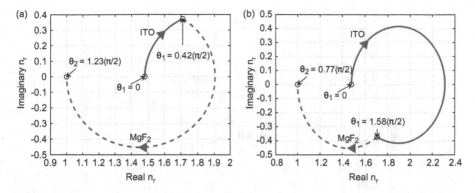

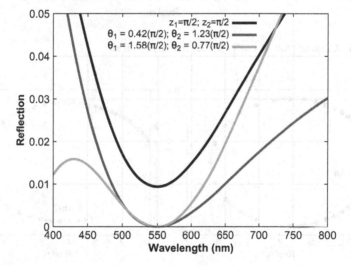

Figure 5.4 Contour plots for the two intersections shown on Figure 5.3. (a) Using the upper intersection. (b) Using the lower intersection.

Figure 5.5 Spectral reflectance plots for the two-layer design using ITO and MgF_2 as shown in Figure 5.4a and b, as well as the quarter-wave design using the same two materials.

However, not all combinations of materials will yield a solution. For example, if we consider LaF_3 with an index of 1.60 instead of Al_2O_3 using quarter-wave films, we would get $n_r = \frac{1.38^2}{1.60^2} 1.48 = 1.1$. This, unfortunately, cannot be corrected by allowing for non-quarter-wave thicknesses. Figure 5.6 shows the effect of varying the two film thicknesses. We can see that the two circles do not intersect.

At this point, it is worth pointing out that the rate of change of phase along a dielectric contour line is not exactly linear. For example, Figure 5.7a shows the contour of a half-wave film, using a substrate index of 1.0 and a film index of 2.5. The dots along the contour represent phase values that are integer multiples of $\pi/40$. We can see that the distance between the points is much smaller on the left side of the plot

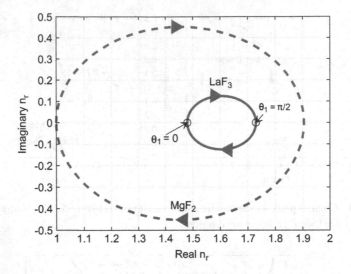

Figure 5.6 Contour plots of LaF$_3$ on silica and MgF$_2$ on air

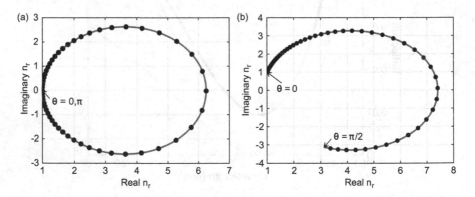

Figure 5.7 Contour plot of a half-wave film (starting from $n_s = 1.0$) and a quarter-wave film (starting from $n_s = 1 + j$) using a film index of 2.5. The dots represent equal increments of phase along the contour. (a) Half-wave film starting from $n_s = 1.0$. (b) Quarter-wave film starting from $n_s = 1 + j$.

than on the right side of the plot. As a consequence, our earlier observation that a quarter-wave-thick film describes half a circle is simply a consequence of the starting and ending points being on the real axis. If the starting point has a complex value, as was the case with the design discussed in Figure 5.4, the length of the contour will not scale linearly. Stated differently, the contour of a quarter-wave film starting from a complex index value will not trace half a circle. An example of this is illustrated in Figure 5.7b.

5.3 THREE-LAYER ANTIREFLECTION DESIGN

One of the drawbacks of the two-layer design compared to the single-layer design is the reduction in the antireflection bandwidth. Figure 5.8 shows the spectral reflection from a single-layer design (as discussed in Chapter 3 using a film index of 1.217 and a thickness of 112.9 nm) and a two-layer design (same as Figure 5.2). The main difference between the two spectra is the reflection bandwidth. It can be clearly seen that the single-layer design has a larger antireflection bandwidth than the two-layer design. This is one of the trade-offs in the two-layer design: by increasing the number of layers, the antireflection bandwidth has been compromised. This arises due to the increase in the resonance quality factor (Q-factor).

One of the commonly used techniques to improve the bandwidth is to a use a third film [1]. But this has to be done without losing the antireflection performance that was achieved at the reference wavelength. One way to do this is by introducing a half-wave-thick film between the two layers. Since a half-wave-thick layer behaves as an absentee film, the performance at the reference wavelength will not be affected. But the performance at all other wavelengths will be modified. This can be easily demonstrated by using a structure such as silica$|HQQL|$air where H is a quarter-wave-thick material of index 1.68, L is a quarter-wave-thick material of index 1.38, and Q is a quarter-wave-thick material of index 2.35. Figure 5.9 shows the reflection spectrum with and without this absentee film. We can clearly see that the spectral bandwidth of the three-layer structure is significantly larger than that of the two-layer structure. In fact, it is slightly larger than even the single-layer design, depending on how you define the spectral width. The basic principle of this three-layer design is discussed in the next section.

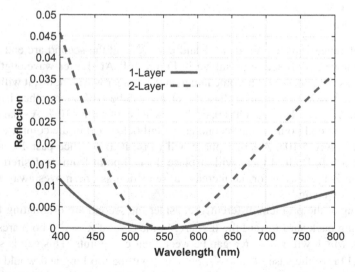

Figure 5.8 Reflectance plots of the one-layer and two-layer designs to demonstrate their effect on the reflection bandwidth.

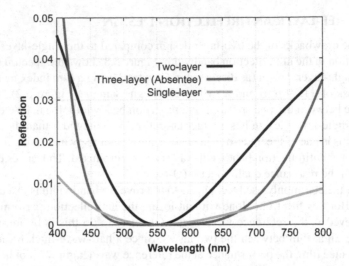

Figure 5.9 Comparison between the one-layer, two-layer, and three-layer designs. The three-layer design uses an absentee layer at the center.

5.4 PRINCIPLES OF THE THREE-LAYER DESIGN USING THE ABSENTEE LAYER

How the absentee layer affects the neighboring wavelengths cannot be easily written down in analytic form, but we can attempt to gain some insights by considering the contours traced by the effective reflectance index. First consider the two-layer structure silica|HL|air where

$$H : n = 1.68$$
$$L : n = 1.38.$$

At the reference wavelength, $\theta_1 = \frac{\pi}{2}$ and $\theta_2 = \frac{\pi}{2}$, and the second arc ends at $n_r = 1.0$ to achieve antireflection as shown in Figure 5.1. At shorter wavelengths, each layer's phase thickness will become larger, and at longer wavelengths, it will become smaller. At $\lambda = \lambda/0.9$ (smaller phase), each arc will be 10% shorter than its original length. At $\lambda = \lambda/1.1$ (larger phase), each arc will be longer by 10%. A smaller phase will cause the end point of the second arc to fall below 1.0 in the complex plane as shown in Figure 5.10a. A larger phase will cause it to advance past 1.0 as shown in Figure 5.10b. Both of these will displace the end point from its desired value of 1.0, which is the reason for the increase in reflection as one moves away from the reference wavelength.

Looking at the problem differently, consider the second arc originating from 1.0 and moving backward toward the first arc. In the case of $\lambda = \frac{\lambda_0}{0.9}$, both arcs will be too short, which will result in a gap between their end points. This gap is shown in Figure 5.11a. In the case of $\lambda = \frac{\lambda_0}{1.1}$, both arcs will be too long and would result in the arcs overrunning each other. These scenarios are shown in Figure 5.11a and b, and the gap is shown by the dotted arrow.

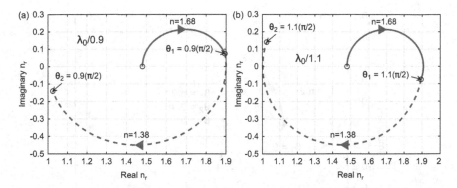

Figure 5.10 Contour of n_r for the two-layer antireflection design for wavelengths slightly longer and slightly shorter than the reference wavelength. (a) $\lambda = \frac{\lambda_0}{0.9}$ ($G = 0.9G_0$). (b) $\lambda = \frac{\lambda_0}{1.1}$ ($G = 1.1G_0$).

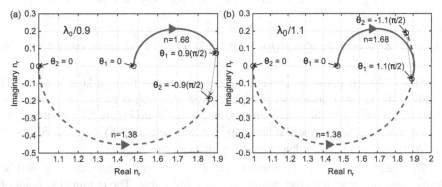

Figure 5.11 Contour of n_r for the two-layer antireflection design for wavelengths slightly longer and slightly shorter than the reference wavelength but with the second arc starting from 1.0 and running backward. (a) $\lambda = \frac{\lambda_0}{0.9}$ ($G = 0.9G_0$). (b) $\lambda = \frac{\lambda_0}{1.1}$ ($G = 1.1G_0$).

The purpose of adding a third layer is to bridge the gap such that it would allow a more reasonable performance to be achieved at $\lambda = \frac{\lambda_0}{0.9}$ and $\lambda = \frac{\lambda_0}{1.1}$ but without affecting the performance at λ_0. This is exactly the function served by the absentee layer.

For example, consider a three-layer-structure silica$|HQQL|$air where

$H : n = 1.68$
$L : n = 1.38$
$Q : n = 2.35$

with $\theta_1 = \frac{\pi}{2}$, $\theta_2 = \pi$, and $\theta_3 = \frac{\pi}{2}$ (at the reference wavelength). Referring to Figure 5.12, at $\lambda = \lambda_0$, the large circle on the right hand side represents the half-wave absentee layer, and the third arc ends at $n_r = 1.0$ as expected. In this case, the absentee layer does not serve any role since the value of n_r will be exactly the same with or without the absentee layer.

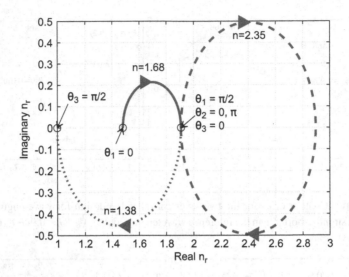

Figure 5.12 Contour for silica|$HQQL$|air at $\lambda = \lambda_0$ ($G = G_0$) with an absentee index of 2.35

At $\lambda = \frac{\lambda_0}{0.9}$, attaching the absentee layer to the end of the first layer's arc would result in the scenario as shown in Figure 5.13a. The gap between the second and third arcs is shown by the dotted arrow. Comparing this to Figure 5.11a (without the absentee layer), we can see that the "gap" has been significantly reduced. Although not perfect, we can expect this to produce a lower reflection at $\lambda = \frac{\lambda_0}{0.9}$.

The scenario at $\lambda = \frac{\lambda_0}{1.1}$ is shown in Figure 5.13b. In this case, each arc will be longer, and the absentee layer would go past a full circle. The termination point of the absentee layer will be 0.1π beyond the starting point. Again, we can see that the "gap" has been significantly reduced compared to Figure 5.11b.

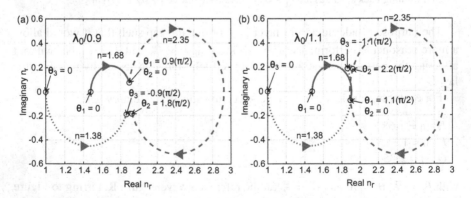

Figure 5.13 Contour for silica|$HQQL$|air for wavelengths slightly longer and slightly shorter than the reference wavelength with an absentee index of 2.35. (a) $\lambda = \frac{\lambda_0}{0.9}$ ($G = 0.9G_0$). (b) $\lambda = \frac{\lambda_0}{1.1}$ ($G = 1.1G_0$).

The refractive index of the absentee layer also plays a very important role. In the above example, we assumed the refractive index of the absentee layer to be 2.35. Because the terminating index of the first layer was about 1.9, this resulted in the absentee layer tracing a circle to the right hand side of the first layer. We can graphically verify that this is a necessary condition to shrink the "gap".

If the index of the absentee layer were lower, for example, 1.5, that would trace a circle to the left of the first arc. We can show that this actually enlarges the gap, making the reflection bandwidth worse. This condition is illustrated in Figure 5.14a and b. The spectral reflectance plots with absentee indices of 2.35 and 1.5 are shown in Figure 5.15.

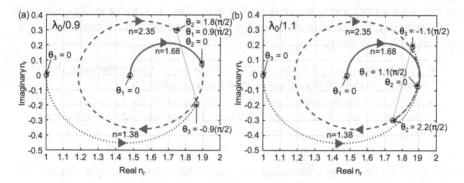

Figure 5.14 Contour for silica$|HQQL|$air for wavelengths slightly longer and slightly shorter than the reference wavelength with an absentee index of 1.5. (a) $\lambda = \frac{\lambda_0}{0.9}$ ($G = 0.9G_0$). (b) $\lambda = \frac{\lambda_0}{1.1}$ ($G = 1.1G_0$).

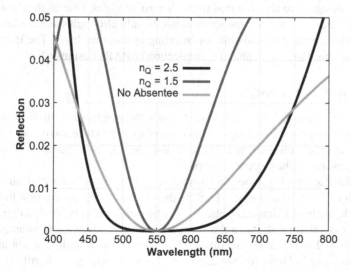

Figure 5.15 Spectral reflectance plots of the three-layer designs using absentee layer indices of 2.35 and 1.5.

This description of how the absentee layer improves the antireflection performance is somewhat qualitative. It does not give us the exact refractive index for the absentee layer, nor the spectral width of the antireflection band. All we can say is that the refractive index of the absentee layer has to be larger than the effective reflectance index of the substrate with the first layer. It is equally important that the index not be too high either. A spacer with too large an index will move the end point in the opposite direction away from 1.0. This will have the effect of reducing the antireflection bandwidth. The optimum value has to be determined to some extent by trial and error. Nevertheless, this description does provide an understanding of how the absentee layer works and an approximate value of its refractive index. On the other hand, instead of the *HL*, if the layer sequence is *LH* (which is rare in most practical scenarios), the absentee index has to be lower to achieve the desired result.

In summary, we can state the following conclusions regarding three-layer *HQQL*-type antireflection designs:

- The absentee layer does not affect the antireflection performance at the reference wavelength regardless of its refractive index.
- In order for the absentee layer to improve the antireflection bandwidth, assuming the layer indices are higher than the substrate index, and the design uses a *HL* sequence, the absentee refractive index must be larger than the effective reflectance index of the substrate and the first layer.

The single-layer coatings discussed in Chapter 3 and the two-layer coatings discussed in this chapter have a single minimum point in their reflection spectrum. This is surrounded by increasing reflection on both sides of the minimum. Since the antireflection point corresponds to a resonant condition, these designs can also be thought of as single-resonance designs. Because their spectral characteristics resemble a "V" shape, the designs are also referred to as "V-coat" designs. One of the drawbacks of the V-coat designs is the narrow spectral bandwidth, although we were later able to significantly expand the bandwidth by inserting an absentee layer. The latter designs are often referred to as broadband antireflection (BBAR) designs [2].

5.5 DOUBLE-V DESIGNS

This is another approach for increasing the antireflection bandwidth. It involves placing two antireflection points adjacent to each other to produce an overall broadening effect on the reflection spectrum. These designs are known as "double-V" designs or "W" designs due to their spectral shape.

Consider the two-layer design illustrated in Figure 5.1 with two quarter-wave films with $n_{f1} = 1.68$ and $n_{f2} = 1.38$. In the double-V design, the first film is half-wave thick, such as silica|HHL|air. Since a half-wave film behaves as an absentee layer, we essentially only have the second film at the reference wavelength. Using $n_{f1} = 1.68$ and $n_{f2} = 1.38$, we obviously cannot achieve antireflection at the reference wavelength. However, we can show that at wavelengths slightly shorter and slightly longer than the reference wavelength, the reflection will decline from its value at the reference wavelength.

Figure 5.16 shows the effective reflectance index contour of this structure at the reference wavelength. We can verify that the end point has a value of $n_r = 1.28$, which would produce a reflection of $\sim 1.5\%$. Now, if we consider two adjacent wavelengths, $\lambda_0/0.8$ and $\lambda_0/1.2$, their corresponding contours are shown in Figure 5.17a and b. It should be apparent that n_r is much closer to n_a, about 1.13, at these two wavelengths. The resulting reflection will be about 0.37%. The calculated reflection spectrum, assuming a reference wavelength of $\lambda_0 = 650$ nm, is shown in Figure 5.18, where we can verify this spectral behavior. While this example illustrates the main principles of the double-V design, we have not yet systematically

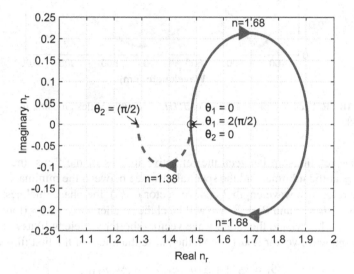

Figure 5.16 Contour for silica|HHL|air, double-V structure at the reference wavelength.

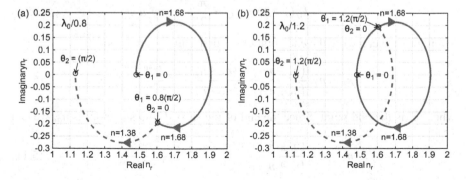

Figure 5.17 Contour for silica|HHL|air double-V structure at wavelengths of $\frac{650}{0.8} = 812.5$ nm and $\frac{650}{1.2} = 541.6$ nm, using $n_{f1} = 1.68$ and $n_{f2} = 1.38$. (a) $\lambda = \frac{\lambda_0}{0.8}$ ($G = 0.8G_0$). (b) $\lambda = \frac{\lambda_0}{1.2}$ ($G = 1.2G_0$).

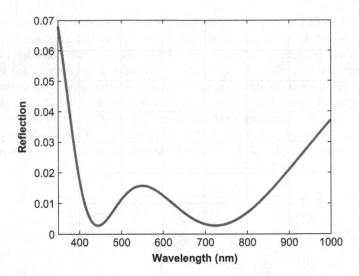

Figure 5.18 Reflectance plot of the silica|HHL|air double-V design using $n_{f1} = 1.68$ and $n_{f2} = 1.38$.

derived the relationship between the refractive indices of the films, the minimum reflectance at the minima, and the spectral distance between the minima.

At the reference wavelength λ_0 (wave vector of k_0), the phase thicknesses of the films will be $\theta_1 = \pi$ and $\theta_2 = \frac{\pi}{2}$. We will label the smaller wave vector (longer wavelength) as $k_0(1 - \Delta)$ and the larger wave vector (shorter wavelength) as $k_0(1 + \Delta)$.

When the wave vector is $k_0(1 + \Delta)$, the phase thickness of the first film will be

$$\theta_1 = k_0(1 + \Delta) n_{f1} a_1 = \pi + \Delta k_0 n_{f1} a_1, \tag{5.8}$$

and for the second film,

$$\theta_2 = k_0(1 + \Delta) n_{f2} a_2 = \frac{\pi}{2} + \Delta k_0 n_{f2} a_2. \tag{5.9}$$

Therefore, we can write the end point of the first film's contour as

$$n_r = n_{f1} \frac{(n_s + n_{f1}) + (n_s - n_{f1}) e^{-2j\Delta k_0 n_{f1} a_1}}{(n_s + n_{f1}) - (n_s - n_{f1}) e^{-2j\Delta k_0 n_{f1} a_1}}, \tag{5.10}$$

where n_r is the end point. Since the end of the second film's contour is n_a, we can write

$$n_a = n_{f2} \frac{(n_r + n_{f2}) - (n_r - n_{f2}) e^{-2jk_0 \Delta n_{f2} a_2}}{(n_r + n_{f2}) + (n_r - n_{f2}) e^{-2jk_0 \Delta n_{f2} a_2}}. \tag{5.11}$$

Note the sign reversal in this equation, which arises from the $\frac{\pi}{2}$ in equation (5.9). We can eliminate n_r from these equations and express it as a single numerical function

$$F(n_{f1}, n_{f2}, \Delta) = n_a. \tag{5.12}$$

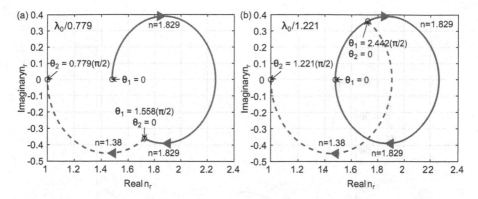

Figure 5.19 Contour for silica|HHL|air double-V design at wavelengths of $\frac{650}{0.779} = 834.4$ nm and $\frac{650}{1.221} = 532.4$ nm, using $n_{f1} = 1.825$ and $n_{f2} = 1.38$. (a) $\lambda = \frac{\lambda_0}{0.779}$ ($G = 0.779G_0$). (b) $\lambda = \frac{\lambda_0}{1.1.223}$ ($G = 1.223G_0$).

Even though it is a single expression, this really represents two equations because it is a complex expression. We can split the real and imaginary parts as

$$\Re \left\{ F \left(n_{f1}, n_{f2}, \Delta \right) \right\} = n_a \tag{5.13}$$

$$\Im \left\{ F \left(n_{f1}, n_{f2}, \Delta \right) \right\} = 0. \tag{5.14}$$

However, there are three unknowns: n_{f1}, n_{f2}, and Δ. If we specify one of these three variables, we can solve for the other two.

For example, if we select $n_{f2} = 1.38$, we can numerically solve this function and get $n_{f1} = 1.829$ and $\Delta = 0.221$. Figure 5.19a and b shows the contour plots at the two wavelengths corresponding to $k_0 (1 - \Delta)$ and $k_0 (1 + \Delta)$. We can see that the end point of the second arc is exactly at n_a in both cases. Figure 5.20 shows the reflectance spectrum for this structure using a reference wavelength of $\lambda_0 = 650$ nm. We can verify that the reflection drops to zero at $\lambda = 532$ nm and $\lambda = 834$ nm.

5.6 ANTIREFLECTION ON A SUBSTRATE THAT ALREADY CONTAINS THIN FILMS

There are instances where we may be faced with designing an antireflection structure on a substrate that already contains a thin film (or films). Since the thickness of the existing films cannot be modified, they have to be considered as a part of the substrate. This situation can be easily handled by calculating the effective reflectance index of the system (substrate plus the existing films) and using that value as the new substrate index. However, since we cannot expect these films to be multiples of quarter waves, the effective reflectance index will not be real-valued. Therefore, this problem requires starting from a substrate whose index is complex. This requires a slightly different mathematical treatment than before.

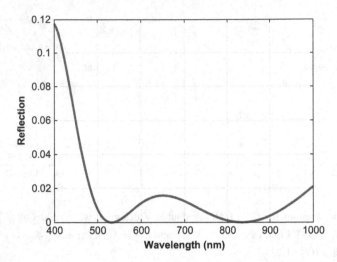

Figure 5.20 Spectral reflectance plot of the numerically solved design discussed in Figure 5.19.

As an example, consider a glass substrate with a 75 nm film of ZnS. We will assume the substrate index is 1.48 and the ZnS index is 2.3 at the reference wavelength of 550 nm. We can calculate the effective reflectance index of the system by using equation (5.6). This results in a value of $n_r = 2.94 - j0.96$. This complex value becomes our substrate refractive index for the purpose of calculating the subsequent antireflection films. We cannot simply take the square root of this index and apply a quarter-wave film. That principle only applies to substrates with real refractive indices.

Using n_r as the substrate index, to achieve antireflection with a single film, we need to satisfy the condition

$$1.0 = n_f \frac{(n_r + n_f) + (n_r - n_f) e^{-j2\theta}}{(n_r + n_f) - (n_r - n_f) e^{-j2\theta}}. \tag{5.15}$$

For a given n_r, we can solve equation (5.15) to obtain values for n_f and θ. It should be remembered that equation (5.15) really consists of two equations – one for the real part and one for the imaginary part. Therefore, we really have two equations with two unknowns. We can write these as

$$\Re \left\{ n_f \frac{(n_r + n_f) + (n_r - n_f) e^{-j2\theta}}{(n_r + n_f) - (n_r - n_f) e^{-j2\theta}} \right\} = 1.0 \tag{5.16}$$

$$\Im \left\{ n_f \frac{(n_r + n_f) + (n_r - n_f) e^{-j2\theta}}{(n_r + n_f) - (n_r - n_f) e^{-j2\theta}} \right\} = 0. \tag{5.17}$$

We can now solve equations (5.16) and (5.17) to find θ and n_f.

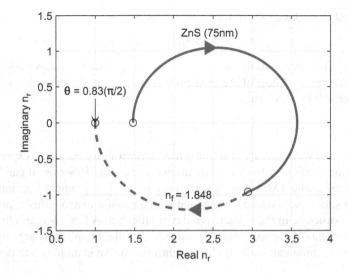

Figure 5.21 Contour plot of the single-layer antireflection design for a silica substrate that already contains a 75 nm ZnS film.

Continuing our example, we can use $n_r = 2.94 - j0.96$ to get $n_f = 1.848$ and $\theta = 0.833 \left(\frac{\pi}{2} \right)$. Using a reference wavelength of 550 nm, this translates to a film thickness of 62 nm. The contour plot of this solution is shown in Figure 5.21.

As we saw in Chapter 3, a single-layer antireflection design may end up with films with unrealistic refractive indices. In such cases, we would have to use the same multilayer design strategies explored in this chapter. For example, consider a Si substrate (index of 3.5 at a wavelength of 1,500 nm) with a 300 nm film of silica (index of 1.5). The resulting value for n_r is $0.697 + j0.390$. This would require a film with an index of 0.44 and a phase thickness of $0.13 \frac{\pi}{2}$. Clearly, this is an unphysical value. However, we can achieve a better solution by using more than one film. This is left as an exercise for the reader.

5.7 STRUCTURED AND GRADIENT-INDEX FILMS

The principles of gradient-index antireflection are based on the idea that a gradual transition in refractive index will produce a smaller reflection compared to a sharp boundary. An example is a thin film whose refractive index gradually increases from n_a to n_s. Since it is not based on interference of waves from two reflection boundaries, in principle, its antireflection performance should be independent of wavelength. However, this is true only to some extent because there is also an additional requirement that the transition region must be much thicker than the wavelength of light in that medium. As a result, there will be a weak wavelength dependence. But this wavelength dependence will not be as strong as with interference films. Sensitivity to the incident angle will also be smaller compared to a regular thin-film antireflection.

The gradient index of the film can be represented as

$$n(z) = n_a + (n_s - n_a) f(z/L), \tag{5.18}$$

where $f(z/L)$ is the function that describes how the refractive index varies from n_a to n_s and L is the thickness of the gradient layer. For example, the simplest case of a linear taper will be represented as

$$f(z/L) = \frac{z}{L}. \tag{5.19}$$

Unlike the case with discrete thin films, it is difficult to formulate an expression for the reflection coefficient from a gradient-index structure. However, it can be easily calculated using the TMM. For this, we have to split the gradient film into a large number of slices and model each segment with a transfer matrix. This approximates a staircase structure in the refractive index profile, but as long as each slice is kept thin, it would closely approximate a continuous profile. Despite the large number of matrices, it is computationally quite straightforward. An example code is shown in Chapter 15.

For example, Figure 5.22a shows the reflection spectrum from a SiO$_2$ substrate with $n_s = 1.48$, with a linearly tapered refractive index from n_a to n_s over a distance L of 1 μm. In the TMM calculation, 1,000 slices were used with each slice thickness of 0.5 nm. We can verify that the reflection values are very low across a wide spectral band. We have assumed the refractive index of the substrate to be a constant. But even with material dispersion, the reflection would still remain low because the performance is dictated by the graduated profile more than the actual values of refractive indices.

Much work has been done in the literature to determine the optimum refractive index profile that yields the lowest reflection [3,4]. Many different profiles, such as gaussian, exponential, and polynomial, have been studied. However, the difference in performance between these profiles is quite small. The main factor is the thickness of the gradient-index layer.

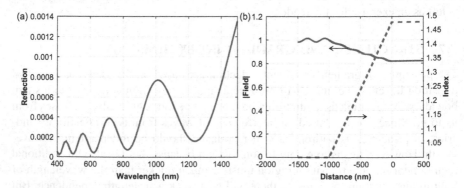

Figure 5.22 Reflection spectrum and field profile for a linearly tapered refractive index profile from $n_s = 1.48$ to $n_a = 1.0$ over a thickness of 1 μm. (a) Reflection spectrum. (b) Field and index distribution at $\lambda = 550$ nm.

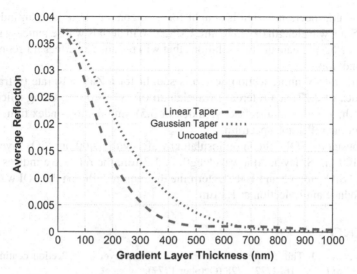

Figure 5.23 Average reflection between 400 and 1,500 nm wavelength as a function of the gradient-index layer thickness for different substrate indices.

Figure 5.23 shows the average reflection value taken between the wavelengths of 400 and 1,500 nm as a function of the gradient-index layer (assuming a linear and Gaussian gradient model). We can see that larger thicknesses result in very low reflection values. As the thickness gets smaller, the average reflection becomes larger and eventually approaches that of the uncoated substrate. A gradient layer that is thinner than about 25 nm behaves nearly the same as a discrete step transition.

Gradient-index layers can be produced by continuously altering the refractive index of a thin film during deposition. This can be done by changing the stoichiometry of the film, or by adjusting the mixture ratio of two different materials. However, the simplest approach is to physically structure the film. This is most commonly done by etching the substrate's surface with cone-shaped structures to synthesize a gradient-refractive-index profile. Such patterns are most commonly found on the eyes of night-flying moths; hence they are also known as moth-eye structures [5]. One of the key advantages of this method is that the antireflection performance can have very broad spectral bandwidths. However, the disadvantage is that these etched features have to be much smaller than the wavelength, requiring patterning at the nanoscale. The principles of converting a tapered structure to a tapered-refractive-index profile is done using the effective medium theory.

5.8 PROBLEMS

1. Using sapphire as the substrate, design a non-quarter-wave antireflection design at $\lambda = 1,550$ nm using Si and SiO_2 as the thin-film materials.
2. For the design in problem 1, determine the optimum material to be used as the absentee layer to improve the antireflection bandwidth.

3. An antireflection design is sought for a silicon substrate with an index of 3.5 at a wavelength of 1,550 nm. Using two materials whose indices are 2.2 and 1.6, calculate the two solutions that will produce a perfect antireflection condition.

4. A double-V antireflection design is sought for a ZnS substrate (refractive index of 2.27) at a reference wavelength of $\lambda_0 = 4.0$ μm. Using silicon as the high-index film (refractive index of 3.5), find the low-index film. Then plot the reflection spectrum.

5. Consider a SOI (silicon on insulator) wafer with a 400 nm SiO_2 layer and a 100 nm Si layer. At a wavelength of 1.3 μm, the refractive indices of Si and SiO_2 are 3.5 and 1.48. Determine the required film on the SOI wafer to produce antireflection at 1.3 μm.

REFERENCES

1. Mouchart, J. Thin film optical coatings. 2: Three-layer antireflection coating theory. *Applied Optics* **16**, 2722–2728 (October 1977).

2. Willey, R. R. Further guidance for broadband antireflection coating design. *Applied Optics* **50**, C274–C278. ISSN: 1530-6984 (October 2014).

3. Grann, E. B., Moharam, M. G. & Pommet, D. A. Optimal design for antireflective tapered two-dimensional subwavelength grating structures. *Journal of the Optical Society of America A* **12**, 333. ISSN: 1084-7529 (Feburary 1995).

4. Kim, K.-H. & Park, Q.-H. Perfect anti-reflection from first principles. *Scientific Reports* **3**, 1062. ISSN: 2045-2322 (December 2013).

5. Perl, E. E., et al. Surface structured optical coatings with near-perfect broadband and wide-angle antireflective properties. *Nano Letters* **14**, 5960–5964. ISSN: 1530-6984 (October 2014).

6 High-Reflection Designs

6.1 EFFECTIVE REFLECTANCE INDEX OF A PERIODIC LAYER

As discussed earlier, the effective reflectance index n_r of the substrate due to N quarter-wave layers is

$$n_{r,N} = n_{f,N}^2 \left(\frac{n_{f,N-2}^2}{n_{f,N-1}^2} \right) \cdots \left(\frac{n_{f,2}^2}{n_{f,3}^2} \right) \left(\frac{n_s}{n_{f,1}^2} \right) \quad \text{if } N \text{ is an even number} \qquad (6.1)$$

and

$$n_{r,N} = \left(\frac{n_{f,N}^2}{n_{f,N-1}^2} \right) \cdots \left(\frac{n_{f,3}^2}{n_{f,2}^2} \right) \left(\frac{n_{f,1}^2}{n_s} \right) \quad \text{if } N \text{ is an odd number.} \qquad (6.2)$$

The reflection coefficient is therefore

$$r = \frac{n_{r,N} - n_a}{n_{r,N} + n_a}. \qquad (6.3)$$

There are two ways in which (6.3) can yield a reflection close to 100% – when $|n_{r,N}| \gg n_a$ or $|n_{r,N}| \ll n_a$. Both of these will result in $r \approx 1$. These conditions can be achieved by making the ratio $\frac{n_{f,N}}{n_{f,N-1}}$ very large, or very small, and also by selecting a large N. In other words, we need a large number of alternating quarter-wave layers with a large refractive index ratio between H and L.

For simplicity, first let's consider two fictitious dispersion-free materials with refractive indices of 2.5 and 1.4, representing them as H and L with quarter-wave thicknesses of 55 and 98.2 nm, respectively, at $\lambda_0 = 550$ nm. For the purposes of this discussion, we will consider these thin films without any substrate to isolate their effects from the substrate. We can examine the effect of adding the layers one by one, up to 20 layers. This is shown in Tables 6.1 and 6.2 for the air$|HLHL\cdots|$air and air$|LHLH\cdots|$air designs, respectively. The spectral reflection plots for the air$|HLHL\cdots|$air structure are shown in Figure 6.1.

The effective reflectance index plots provide insights into the mechanism. Let's consider a simple air$|HLHL|$air structure with $H = 2.5$ and $L = 1.4$. The contour of n_r is shown in Figure 6.2a. It traces a constantly expanding spiral. The starting point has an effective index of 1.0 (since we assume the substrate to be air). After a quarter layer of the H material, the trace moves toward the right and terminates at 6.25. After the next L layer, the trace moves back toward lower values and terminates at 0.3. After each additional layer, the effective index progressively moves toward higher and lower values alternatingly.

Notice that the increase in reflectivity is not necessarily monotonic. For example, the peak reflection from air$|H|$air is 52%, but the peak reflection from air$|HL|$air is only 27.3%. We can see from this example that beginning and ending the structure with the H layer produces the highest reflection. However, for large numbers of

Table 6.1

Effective Reflectance Index and Reflection as a Function of Increasing Layers in an Air|$HLHL\cdots$|Air Design

Structure	n_r	Reflection
Air|H|air	$\frac{2.5^2}{1.0} = 6.25$	$\left\|\frac{1.0-6.250}{1.0+6.250}\right\|^2 = 52.0\%$
Air|HL|air	$\frac{1.4^2}{6.25} = 0.313$	$\left\|\frac{1.0-0.313}{1.0+0.313}\right\|^2 = 27.3\%$
Air|HLH|air	$\frac{2.5^2}{0.313} = 19.92$	$\left\|\frac{1.0-19.92}{1.0+19.92}\right\|^2 = 81.8\%$
Air|$HLHL$|air	$\frac{1.4^2}{19.92} = 0.098$	$\left\|\frac{1.0-0.098}{1.0+0.098}\right\|^2 = 67.4\%$
Air|$HLHLH$|air	$\frac{2.5^2}{0.098} = 63.55$	$\left\|\frac{1.0-63.55}{1.0+63.55}\right\|^2 = 93.9\%$
Air|$(HL)^{10}$|air	$\left(\frac{1.4^2}{2.5^2}\right)^{10} 1.0 = 9.2\times10^{-6}$	$\left\|\frac{1.0-9.2\times10^{-6}}{1.0+9.2\times10^{-6}}\right\|^2 = 99.99\%$

Table 6.2

Effective Reflectance Index and Reflection as a Function of Increasing Layers in an Air|$HLHL\cdots$|Air Design

Structure	n_r	Reflection
Air|L|air	$\frac{1.4^2}{1.0} = 1.96$	$\left\|\frac{1.0-1.96}{1.0+1.96}\right\|^2 = 10.5\%$
Air|LH|air	$\frac{2.5^2}{1.96} = 3.18$	$\left\|\frac{1.0-3.18}{1.0+3.18}\right\|^2 = 27.3\%$
Air|LHL|air	$\frac{1.4^2}{3.18} = 0.61$	$\left\|\frac{1.0-0.61}{1.0+0.61}\right\|^2 = 5.7\%$
Air|$LHLH$|air	$\frac{2.5^2}{0.61} = 10.17$	$\left\|\frac{1.0-10.17}{1.0+10.17}\right\|^2 = 67.4\%$
Air|$LHLHL$|air	$\frac{1.4^2}{10.17} = 0.192$	$\left\|\frac{1.0-0.192}{1.0+0.192}\right\|^2 = 45.8\%$
Air|$(LH)^{10}$|air	$\left(\frac{2.5^2}{1.4^2}\right)^{10} 1.0 = 1.1\times10^{5}$	$\left\|\frac{1.0-1.1\times10^5}{1.0+1.1\times10^5}\right\|^2 = 99.99\%$

layers, both structures converge toward the same high-reflection value at the reference wavelength.

The reason for the difference between the air|$HLHLH$|air and air|$LHLHL$|air can also be understood by examining the contours of n_r of these structures. In the air|$HLHLH$|air structure, the spiral starts with the first layer and continues to expand, as shown in Figure 6.2a. Figure 6.2b shows the corresponding plot for air|$LHLHL$|air. Here, we can see that the first L and H produce consecutive arcs in the same direction because both L and H have indices higher than the assumed substrate index of 1.0. This gives us a guideline on which layer should be first. If the substrate index is closer to L, then it is better to start with H. If the substrate index is closer to H, then it is better to start the first layer with L.

We can also see that in order to obtain the highest reflection with the fewest layers, we need to have an H material with the highest index and an L material with the

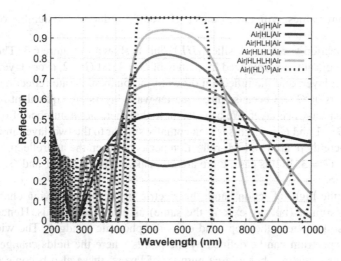

Figure 6.1 Spectral reflection for increasing layers for air|$HLHL\cdots$|air design.

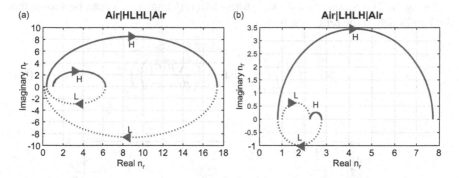

Figure 6.2 Complex effective reflectance index contours. (a) Air|$HLHL\cdots$|air. (b) Air|$LHLH\cdots$|air.

lowest index. In the visible spectrum, these materials are typically MgF$_2$, which has an index of 1.38, and TiO$_2$, which has an index of 2.5. In the infrared spectrum, one could use MgF$_2$ and Ge (which has an index of 4.0). It is also useful to estimate how many layer pairs one needs to achieve 99.9% reflection. For the MgF$_2$/TiO$_2$ pair, we can verify that we would need seven pairs of (HL) to achieve 99.9%. For MgF$_2$/Ge, we would need only four pairs. With a weak refractive index ratio, such as MgF$_2$/LaF$_2$, it would take 16 pairs to reach 99.9%. These results are for a freestanding film structure without accounting for the substrate. The substrate will, of course, change the performance depending on its refractive index.

Also, in contrast to the antireflection designs, the high-reflection designs do not require the end point of n_r to reach a specific value (which is n_a). It just needs to be as high or as small as possible. Therefore, refractive indices of the high/low pairs can have any values, with large ratios preferred for reducing the number of required layers.

The reflection spectrum of silica$|(HL)^{10}|$air is shown in Figure 6.3. The peak reflection occurs at the normalized frequency of $G = 1$. At $G = 2$, every layer becomes an absentee layer, and the reflection drops to the plain substrate reflection. Another peak appears at $G = 3$ because three quarter-wave layers are equivalent in phase to a single quarter-wave layer. We can also note that there are a number of oscillations between $G = 1$ and $G = 3$. This is conceptually similar to the two-layer antireflection design where there are two high-reflection peaks between the antireflection points at $G = 1$ and $G = 3$. In this case, there are 20 peaks between $G = 1$ and $G = 3$ due to the 20 layers.

Within the high-reflection band, the electric field has a decaying characteristic, somewhat similar (but not exactly the same) as evanescent fields. Hence, this region is also known as the stop band or as the photonic bandgap. The width of the reflection spectrum can be defined by the points where the fields change from decaying to propagating. For a large number of layers, these also become the points where the reflection drops to zero. We will examine these aspects in greater detail in Chapter 7.

Using the Herpin equivalent index formulation (Chapter 7), it can be shown that the bandwidth of the reflection is

$$\Delta G = \left(\frac{2}{\pi}\right)\left(\pi - 2\sin^{-1}\left(\frac{2\sqrt{n_1 n_2}}{n_1 + n_2}\right)\right), \tag{6.4}$$

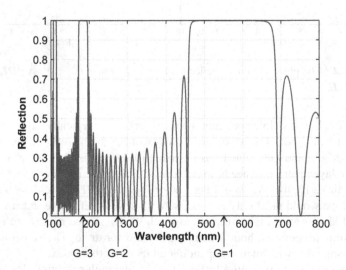

Figure 6.3 Reflection spectrum of silica$|(HL)^{10}|$air showing the normalized frequencies $G = 1, 2, 3$.

which in terms of wavelength becomes

$$\Delta\lambda = \lambda_0 \left[\frac{\pi/2}{\sin^{-1}\left(\frac{2\sqrt{n_1 n_2}}{n_1 + n_2}\right)} - \frac{\pi/2}{\pi - \sin^{-1}\left(\frac{2\sqrt{n_1 n_2}}{n_1 + n_2}\right)} \right]. \tag{6.5}$$

Using our values of $n_H = 2.5$ and $n_L = 1.4$, we can calculate $\Delta G = 0.364$, which corresponds to $\Delta\lambda = 207$ nm using a reference wavelength of $\lambda_0 = 550$ nm.

It should be noted that we have considered the refractive index to be a constant in all of these analyses. In reality, material dispersion will cause some differences, especially over a large spectral range.

6.2 SYMMETRIC UNIT CELL

For a better mathematical treatment, it is useful to define the repeating period of this structure in terms of a symmetric unit cell. This unit cell can be either $\left(\frac{H}{2}L\frac{H}{2}\right)$ or $\left(\frac{L}{2}H\frac{L}{2}\right)$. The main difference between $\left(\frac{H}{2}L\frac{H}{2}\right)^N$ or $\left(\frac{L}{2}H\frac{L}{2}\right)^N$ and $(HL)^N$ is the outer layer. $\left(\frac{H}{2}L\frac{H}{2}\right)^N$ and $\left(\frac{L}{2}H\frac{L}{2}\right)^N$ terminate with one-eighth-wave films of the same material, whereas $(HL)^N$ terminates with quarter-wave films of the opposite materials. Nevertheless, this three-layer unit cell has a clearer mathematical interpretation as we will see in Chapter 7.

6.3 HIGH-REFLECTION DESIGNS WITH SYMMETRIC UNIT CELLS

If the repeating unit cell is $\left(\frac{H}{2}L\frac{H}{2}\right)^{10}$ where H has an index of 2.5 and L has an index of 1.4, with no substrate, the resulting reflection spectrum will show a central high-reflection band with a high transmission band on the longer-wavelength side of the spectrum. The opposite is true with the $\left(\frac{L}{2}H\frac{L}{2}\right)^{10}$ structure. The shorter-wavelength side will have a high transmission band. As a result, $\left(\frac{H}{2}L\frac{H}{2}\right)$ becomes the unit cell for the construction of a long-pass filter, while $\left(\frac{L}{2}H\frac{L}{2}\right)$ will be the unit cell for a short-pass filter, as shown in Figure 6.4. Of course, these will require additional modifications to improve the steepness of the transition and the flatness of the pass band. The mathematical basis for this analysis will be discussed in Chapter 8.

6.4 BROADBAND REFLECTORS

The width of the reflection band is limited by the high and low refractive indices used in the periodic stack. In the visible range, the highest refractive index is about 2.5, and the lowest is about 1.38. From equation (6.5), we can calculate that the maximum bandwidth that we can realistically achieve is about 200 nm.

One possible technique to extend the width of the reflection band is to use two periodic stacks. However, this has to be done with caution, because in general, the performance of a combined structure is not a simple sum or product of the performances of the individual structures in isolation.

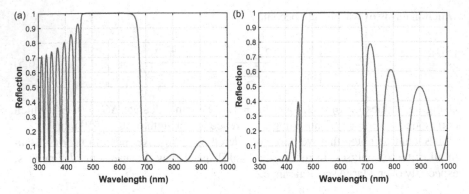

Figure 6.4 Reflection spectrum of two periodic stacks to illustrate their differences outside their central reflection band. (a) $\left(\frac{H}{2}L\frac{H}{2}\right)^{10}$. (b) $\left(\frac{L}{2}H\frac{L}{2}\right)^{10}$.

For this example, we will consider two stacks: $\left(\frac{H}{2}L\frac{H}{2}\right)^{10}$ and $\left(\frac{L}{2}H\frac{L}{2}\right)^{10}$. However, the $\left(\frac{H}{2}L\frac{H}{2}\right)$ stack is designed for a shorter reference wavelength, and the $\left(\frac{L}{2}H\frac{L}{2}\right)$ stack is designed for a longer wavelength, so that the combined effect can exhibit a side-by-side combination of reflectors to increase the overall reflection band. We will write this combined stack as silica$|\left(\frac{H}{2}L\frac{H}{2}\right)^{10}\left(\frac{l}{2}h\frac{l}{2}\right)^{10}|$air where H and L are quarter waves at $\lambda_0 = 550$ nm and h and l are quarter waves at $\lambda_0 = 730$ nm. The individual performance of these stacks without a substrate is shown in Figure 6.5.

The combined effect is shown in Figure 6.6. We can note the net effect is nearly that of a simple sum of the two reflection spectra from Figure 6.5. This has been

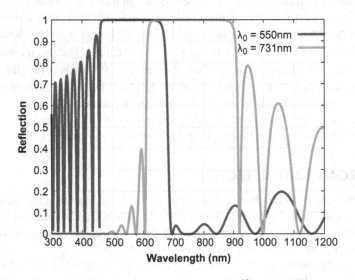

Figure 6.5 Reflection spectrum of $\left(\frac{H}{2}L\frac{H}{2}\right)^{10}$ and $\left(\frac{l}{2}h\frac{l}{2}\right)^{10}$ with different reference wavelengths.

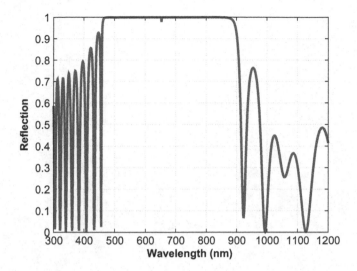

Figure 6.6 Combined stack of $\mathrm{air}\left|\left(\frac{H}{2}L\frac{H}{2}\right)^{10}\left(\frac{l}{2}h\frac{l}{2}\right)^{10}\right|\mathrm{air}$ from Figure 6.5.

possible only because the high-reflectivity portion of one spectrum overlaps with the low-reflectivity portion of the other spectrum. In regions where the high reflection of both regions overlap, there is a potential of creating resonant modes that can create high transmission peaks inside the high-reflection band. We can observe one such transmission peak at the center of the spectrum. This is due to the small overlap between the high-reflection bands in Figure 6.5 near 650 nm.

7 Herpin Equivalence Principle

7.1 BASIC PRINCIPLES

The basic idea behind the Herpin equivalence principle is that a layer sandwiched between two other identical layers, such as (ABA), can be described as having an equivalent index η and an equivalent phase thickness θ as illustrated in Figure 7.1. Therefore, repeating periods of this three-layer unit cell, such as $(ABA)^N$, will also have the same equivalent index η with a phase thickness of $N\theta$. In other words, adding more unit cells only changes the effective phase length of the structure. This is a very useful mathematical construction because treating a periodic structure as if it had a uniform refractive index allows us to use it as a building block to synthesize more complex structures.

7.2 PREVIEW EXAMPLE

Before deriving the Herpin's equivalence principle, its usefulness may be best illustrated through an example. (We will derive these expressions from first principles a bit later.)

Consider a periodic stack $\left(\frac{H}{2}L\frac{H}{2}\right)^5$ with $n_H = 2.5$ and $n_L = 1.5$ on a substrate with refractive index $n_s = 1.5$. At the reference wavelength, using the Herpin equivalent method, the phase thickness of the unit cell can be shown to be

$$\theta = \cos^{-1}\left\{-\frac{1}{2}\left(\frac{n_H}{n_L} + \frac{n_L}{n_H}\right)\right\} \tag{7.1}$$

$$= \cos^{-1}\{-1.13\} \tag{7.2}$$

$$= \pi - j0.51. \tag{7.3}$$

A complex value for the phase thickness may appear a bit strange, but it is a feature of the stop band (or photonic bandgap) of a periodic structure. Additionally, the equivalent index of the unit cell at the reference wavelength turns out to be a purely imaginary number:

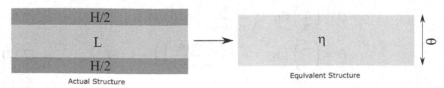

Actual Structure

Equivalent Structure

Figure 7.1 Equivalent index representation of a three-layer unit cell $\left(\frac{H}{2}L\frac{H}{2}\right)$.

$$\eta = jn_H \tag{7.4}$$

$$= j2.5. \tag{7.5}$$

Therefore, the total phase thickness of five repeating unit cells will be

$$N\theta = 5\pi - j2.55. \tag{7.6}$$

These may seem like abstract quantities, but their usefulness lies in how they can be used to calculate results. For example, assuming this film stack is on a substrate with a refractive index of n_s, we can calculate its effective reflectance index by using the principles developed in the previous chapters. That is, the effective reflectance index can be written as

$$n_r = \eta \frac{(n_s + \eta) + (n_s - \eta) e^{-j2N\theta}}{(n_s + \eta) - (n_s - \eta) e^{-j2N\theta}}. \tag{7.7}$$

From this, using $n_s = 1.5$, we can get $n_r = 0.026 + j2.486$. Therefore, the reflection can be calculated to be

$$R = \left| \frac{n_r - n_a}{n_r + n_a} \right|^2 = 0.985. \tag{7.8}$$

Furthermore, if n_a and n_s are assumed to have the same refractive index as n_H (the outer layers in the trilayer unit cell), the reflection at the reference wavelength becomes a simple expression

$$R = |\tan N\theta|^2 \tag{7.9}$$

$$= 0.976. \tag{7.10}$$

We can see that the incident medium and the substrate have minimal impact on the overall reflection spectrum because both reflection coefficients (equations 7.8 and 7.10) are very close to each other. We can also see that R at the reference wavelength will increase as N is increased. This arises due to the imaginary part of θ.

The width of the reflection band is defined as when the equivalent phase thickness of the unit cell θ switches from being complex to real. This also coincides with the equivalent phase of the unit cell becoming equal to π (or alternatively, becoming an absentee layer). Unlike in a single layer where the absentee condition occurs at half the reference wavelength (or at two times the normalized frequency, $G = 2$), in this case, it occurs at two frequencies on either side of the normalized frequency. These frequencies can be derived to be

$$G = \left(\frac{2}{\pi}\right) \sin^{-1} \left(\frac{2\sqrt{n_1 n_2}}{n_1 + n_2}\right) = 0.839 \tag{7.11}$$

$$G = \left(\frac{2}{\pi}\right) \left(\pi - \sin^{-1} \left(\frac{2\sqrt{n_1 n_2}}{n_1 + n_2}\right)\right) = 1.161. \tag{7.12}$$

If the reference wavelength is $\lambda_0 = 550$ nm, the reflection band spans between $\frac{550}{0.839} = 655.4$ nm and $\frac{550}{1.161} = 473.8$ nm for a bandwidth of 181.6 nm.

The effective reflectance index n_r can also be calculated at these wavelengths, and this can be shown to be

$$n_r = \frac{n_s}{1 + j2n_sN\frac{(1-n_2/n_1)}{\sqrt{n_1 n_2}}}, \tag{7.13}$$

for one edge of the stop band and

$$n_r = n_s + j2N\frac{(n_1 - n_2)}{\sqrt{n_1 n_2}} \tag{7.14}$$

for the other edge of the stop band. Assuming the substrate index $n_s = 1.5$, this results in $n_r = 0.1415 - j0.4385$, from which we can calculate the reflection as

$$R = \left| \frac{n_r - n_a}{n_r + n_a} \right|^2 = 0.62 \tag{7.15}$$

for one edge and $n_r = 1.5 + j12.91$ for the other edge, resulting in

$$R = \left| \frac{n_r - n_a}{n_r + n_a} \right|^2 = 0.97. \tag{7.16}$$

The numerically calculated spectrum of this $\left(\frac{H}{2}L\frac{H}{2}\right)^5$ structure is shown in Figure 7.2. We should be able to verify that all of the analytically calculated values are consistent with this spectrum.

The Herpin's equivalence model also allows us to calculate other aspects of this structure, such as the zero reflection points outside the main reflection band, as well as provides guidelines on how to reduce the oscillation peaks in these areas.

This example is meant simply to illustrate the usefulness of the equivalent index principle. In the next sections, we will systematically derive these equations from first principles.

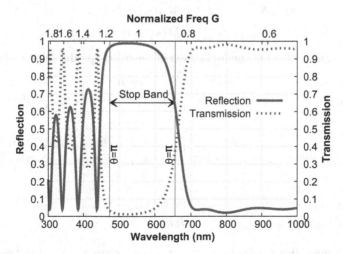

Figure 7.2 Reflection and transmission spectra calculated using TMM for $\left(\frac{H}{2}L\frac{H}{2}\right)^5$ with $n_H = 2.5$, $n_L = 1.5$, and $n_s = 1.5$.

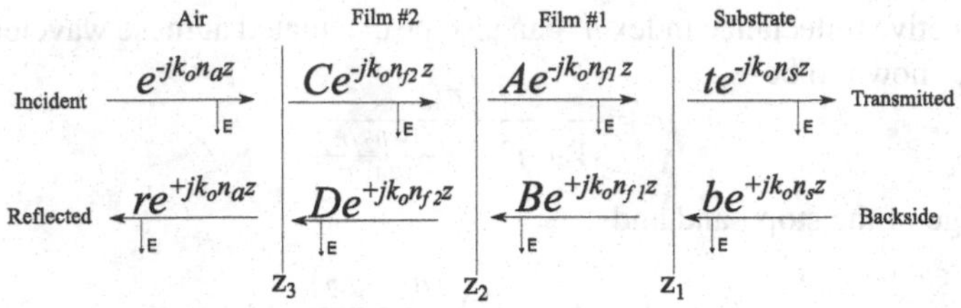

Figure 7.3 Normally incident fields with two thin films on a substrate.

7.3 TRILAYER UNIT CELL

For a two-layer system as shown in Figure 7.3, in Chapter 4, we derived the characteristic transfer matrix of the system as

$$
\begin{bmatrix} 1 \\ r \end{bmatrix} = [M(n_a, z_3)]^{-1} \underbrace{M(n_{f2}, z_3) [M(n_{f2}, z_2)]^{-1}}_{N(n_{f2}, z_2, z_3)}
$$

$$
\times \underbrace{M(n_{f1}, z_2) [M(n_{f1}, z_1)]^{-1}}_{N(n_{f1}, z_1, z_2)} M(n_s, z_1) \begin{bmatrix} t \\ 0 \end{bmatrix}. \tag{7.17}
$$

As we did in Chapter 4, we can group these matrices based on which layers they represent. We can identify that the first matrix on the left $[M(n_a, z_3)]^{-1}$ arises from the outside medium (air) and the last matrix on the right $M(n_s, z_1)$ arises due to the substrate. We can attribute the other matrices to each film as indicated by the underbraces in equation (7.17):

$$
N_2 = N(n_{f2}, z_2, z_3) = M(n_2, z_3) [M(n_2, z_2)]^{-1}, \tag{7.18}
$$

$$
N_1 = N(n_{f1}, z_1, z_2) = M(n_1, z_2) [M(n_1, z_1)]^{-1}. \tag{7.19}
$$

For N_1, we can carry out the multiplication of the matrix product algebraically

$$
N_1 = \begin{bmatrix} e^{-jk_0 n_1 z_2} & e^{+jk_0 n_1 z_2} \\ jk_0 n_1 e^{-jk_0 n_1 z_2} & -jk_0 n_1 e^{+jk_0 n_1 z_2} \end{bmatrix} \begin{bmatrix} e^{-jk_0 n_1 z_1} & e^{+jk_0 n_1 z_1} \\ jk_0 n_1 e^{-jk_0 n_1 z_1} & -jk_0 n_1 e^{+jk_0 n_1 z_1} \end{bmatrix}^{-1}
$$

$$
= \begin{bmatrix} e^{-jk_0 n_1 z_2} & e^{+jk_0 n_1 z_2} \\ jk_0 n_1 e^{-jk_0 n_1 z_2} & -jk_0 n_1 e^{+jk_0 n_1 z_2} \end{bmatrix} \begin{bmatrix} -jk_0 n_1 e^{+jk_0 n_1 z_1} & -e^{+jk_0 n_1 z_1} \\ -jk_0 n_1 e^{-jk_0 n_1 z_1} & e^{-jk_0 n_1 z_1} \end{bmatrix} \frac{1}{-j2k_0 n_1}
$$

$$
= \begin{bmatrix} \cos(k_0 n_1 (z_2 - z_1)) & -\frac{1}{k_0 n_1} \sin(k_0 n_1 (z_2 - z_1)) \\ k_0 n_1 \sin(k_0 n_1 (z_2 - z_1)) & \cos(k_0 n_1 (z_2 - z_1)) \end{bmatrix}. \tag{7.20}
$$

We can recognize that the term $k_0 n_1 (z_2 - z_1)$ is the phase thickness of layer #1. We can abbreviate this as

$$
\delta_1 = k_1 (z_2 - z_1), \tag{7.21}
$$

where

$$k_1 = k_0 n_1. \tag{7.22}$$

As a result, we can write the matrix N_1 as

$$N_1 = \begin{bmatrix} \cos\delta_1 & -\frac{1}{k_1}\sin\delta_1 \\ k_1\sin\delta_1 & \cos\delta_1 \end{bmatrix}. \tag{7.23}$$

This matrix has two special properties: (1) the diagonal elements are equal to one another, and (2) the determinant of the matrix is equal to 1. Our goal is to come up with a sequence of films $N_1 N_2 N_3 \dots$ that would share these same properties as a single film. In other words, we want the sequence $N_1 N_2 N_3 \dots$ to also have equal diagonal elements and a determinant value of 1. Since $det(AB) = det(A)\,det(B)$, the determinant of $N_1 N_2 N_3 \dots$ will always be 1 regardless of the exact sequence. However, the diagonals of $N_1 N_2 N_3 \dots$ will not be equal unless the sequence of the matrices is symmetric. Examples of symmetric sequences include $N_1 N_2 N_1$ and $N_1 N_2 N_3 N_2 N_1$.

Let's consider the symmetric sequence $N_1 N_2 N_1$. This will result in a characteristic transfer matrix that can be expanded as

$$N_1 N_2 N_1 = \begin{bmatrix} \cos\delta_1 & -\frac{1}{k_1}\sin\delta_1 \\ k_1\sin\delta_1 & \cos\delta_1 \end{bmatrix} \begin{bmatrix} \cos\delta_2 & -\frac{1}{k_2}\sin\delta_2 \\ k_2\sin\delta_2 & \cos\delta_2 \end{bmatrix}$$
$$\begin{bmatrix} \cos\delta_1 & -\frac{1}{k_1}\sin\delta_1 \\ k_1\sin\delta_1 & \cos\delta_1 \end{bmatrix}, \tag{7.24}$$

where

$$\delta_2 = k_2(z_3 - z_2) \tag{7.25}$$
$$k_2 = k_0 n_2. \tag{7.26}$$

Because the product $N_1 N_2 N_1$ will have equal diagonal elements, it can be generically written as

$$N_1 N_2 N_1 = \begin{bmatrix} a & b \\ c & a \end{bmatrix}. \tag{7.27}$$

By drawing parallels with equation (7.23), we can set the diagonal elements to be $\cos\theta$ (where θ is yet to be defined):

$$N_1 N_2 N_1 = \begin{bmatrix} \cos\theta & b \\ c & \cos\theta \end{bmatrix}. \tag{7.28}$$

Furthermore, since the determinant is known to be equal to 1.0, we can get

$$\cos^2\theta - bc = 1, \tag{7.29}$$

which results in

$$bc = \cos^2\theta - 1 = -\sin^2\theta. \tag{7.30}$$

With the goal of seeking a matrix that looks like the characteristic matrix of a single layer as in equation (7.23), we can set

$$b = -\frac{1}{q}\sin\theta . \tag{7.31}$$

This will result in

$$c = q\sin\theta. \tag{7.32}$$

Therefore, our three-layer sequence becomes

$$N_H = N_1 N_2 N_1 = \begin{bmatrix} \cos\theta & -\frac{1}{q}\sin\theta \\ q\sin\theta & \cos\theta \end{bmatrix}, \tag{7.33}$$

where the subscript H stands for the Herpin equivalent matrix.

For this sequence of three layers, the full transfer matrix equation becomes

$$\begin{bmatrix} 1 \\ r \end{bmatrix} = [M(n_a, c)]^{-1} \, N_H \, M(n_s, a) \begin{bmatrix} t \\ 0 \end{bmatrix}. \tag{7.34}$$

What is even more useful about this equivalent matrix is that repeating this three-layer sequence N times simply results in a matrix with a phase thickness of $N\theta$. That is,

$$(N_H)^N = \begin{bmatrix} \cos\theta & -\frac{1}{q}\sin\theta \\ q\sin\theta & \cos\theta \end{bmatrix}^N \tag{7.35}$$

$$= \begin{bmatrix} \cos N\theta & -\frac{1}{q}\sin N\theta \\ q\sin N\theta & \cos N\theta \end{bmatrix}. \tag{7.36}$$

This should not be a surprising result because repeating a single layer N times should produce a single thicker film with no other difference, except in this case we are actually repeating a unit cell that consists of three layers.

Though it may seem a bit long winded, this is a powerful result because it allows us to consider the three-layer stack $N_1 N_2 N_1$ as an equivalent single layer with a phase thickness of θ and a wave vector of q. Comparing equation (7.33) with (7.23), the term $\eta = \frac{q}{k_0}$ can be considered as the equivalent index of the stack and is known as the Herpin equivalent index [1,2]. More importantly, since the three layers are represented as a single layer, repeating patterns of the same three layers can be treated very simply as a single thicker layer. In other words, repeating the equivalent single layer N times will produce a layer with a phase thickness of $N\theta$ with the same equivalent index $\eta = \frac{q}{k_0}$. This is illustrated in Figure 7.4.

What remains now is to find the expressions relating the phase thickness θ and the equivalent index η to the actual physical parameters of the unit cell. By multiplying out the individual elements of N_1, N_2, and N_1, we can get the following expressions for the matrix elements N_H:

$$\cos\theta = \cos 2\delta_1 \cos\delta_2 - \frac{1}{2}\left(\frac{k_1}{k_2} + \frac{k_2}{k_1}\right)\sin 2\delta_1 \sin\delta_2 \tag{7.37}$$

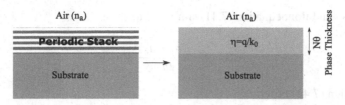

Figure 7.4 Representation of the multilayer film stack as an equivalent film sandwiched between the substrate and air.

$$\frac{1}{q}\sin\theta = \frac{1}{k_1}\left\{\sin\delta_2\left[\frac{1}{2}\left(\frac{k_1}{k_2}+\frac{k_2}{k_1}\right)\cos2\delta_1 + \frac{1}{2}\left(\frac{k_1}{k_2}-\frac{k_2}{k_1}\right)\right] + \sin2\delta_1\cos\delta_2\right\}$$
(7.38)

$$q\sin\theta = k_1\left\{\sin\delta_2\left[\frac{1}{2}\left(\frac{k_1}{k_2}+\frac{k_2}{k_1}\right)\cos2\delta_1 - \frac{1}{2}\left(\frac{k_1}{k_2}-\frac{k_2}{k_1}\right)\right] + \sin2\delta_1\cos\delta_2\right\}.$$
(7.39)

Dividing equation (7.39) by (7.38), we can get an expression for q:

$$q = k_1\sqrt{\frac{\sin\delta_2\left[\frac{1}{2}\left(\frac{k_1}{k_2}+\frac{k_2}{k_1}\right)\cos2\delta_1 - \frac{1}{2}\left(\frac{k_1}{k_2}-\frac{k_2}{k_1}\right)\right] + \sin2\delta_1\cos\delta_2}{\sin\delta_2\left[\frac{1}{2}\left(\frac{k_1}{k_2}+\frac{k_2}{k_1}\right)\cos2\delta_1 + \frac{1}{2}\left(\frac{k_1}{k_2}-\frac{k_2}{k_1}\right)\right] + \sin2\delta_1\cos\delta_2}}.$$
(7.40)

These can be rewritten in terms of the film indices n_1 and n_2, and the free space wave vector k_0, resulting in

$$\cos\theta = \cos2\delta_1\cos\delta_2 - \frac{1}{2}\left(\frac{n_1}{n_2}+\frac{n_2}{n_1}\right)\sin2\delta_1\sin\delta_2$$
(7.41)

$$\eta = \frac{q}{k_0} = n_1\sqrt{\frac{\sin\delta_2\left[\frac{1}{2}\left(\frac{n_1}{n_2}+\frac{n_2}{n_1}\right)\cos2\delta_1 - \frac{1}{2}\left(\frac{n_1}{n_2}-\frac{n_2}{n_1}\right)\right] + \sin2\delta_1\cos\delta_2}{\sin\delta_2\left[\frac{1}{2}\left(\frac{n_1}{n_2}+\frac{n_2}{n_1}\right)\cos2\delta_1 + \frac{1}{2}\left(\frac{n_1}{n_2}-\frac{n_2}{n_1}\right)\right] + \sin2\delta_1\cos\delta_2}}.$$
(7.42)

7.4 TRILAYER UNIT CELL WITH $\delta_2 = 2\delta_1$

In the previous description of $(N_1N_2N_1)^N$, the sequence of phase thicknesses of the individual layers will be

$$\delta_1|\delta_2|2\delta_1|\delta_2|2\delta_1|\delta_2|2\delta_1|\delta_2|\ldots|\delta_1.$$

In other words, except for the outer two layers, the structure consists of repeating phase thicknesses of δ_2 and $2\delta_1$. A special case arises when the phase thicknesses of the two layers are made identical. That is,

$$2\delta_1 = \delta_2.$$
(7.43)

Using this condition, equation (7.41) simplifies to

$$\cos\theta = 1 - \frac{(n_1+n_2)^2}{2n_1 n_2}\sin^2\delta_2, \tag{7.44}$$

and equation (7.42) becomes

$$\eta = n_1 \sqrt{\frac{\cos\delta_2 - (n_1-n_2)/(n_1+n_2)}{\cos\delta_2 + (n_1-n_2)/(n_1+n_2)}}. \tag{7.45}$$

It is useful to rewrite equations (7.44) and (7.45) in terms of the normalized frequency G. We will define G such that the phase of the central layer will be equal to $\frac{\pi}{2}$ when $G=1$. Therefore, δ_2 can be written in terms of G as $\delta_2 = \frac{G\pi}{2}$. As a result, equations (7.44) and (7.45) can be written in terms of G as

$$\cos\theta = 1 - \frac{(n_1+n_2)^2}{2n_1 n_2}\sin^2\left(\frac{G\pi}{2}\right) \tag{7.46}$$

$$\eta = \frac{q}{k_0} = n_1\sqrt{\frac{\cos\left(\frac{G\pi}{2}\right) - (n_1-n_2)/(n_1+n_2)}{\cos\left(\frac{G\pi}{2}\right) + (n_1-n_2)/(n_1+n_2)}}. \tag{7.47}$$

Equations (7.46) and (7.47) are the general expressions, but they simplify even further at the reference wavelength ($G=1$). From equation (7.46), we can simplify θ as

$$\theta|_{G=1} = \cos^{-1}\left[-\frac{1}{2}\left(\frac{n_1}{n_2} + \frac{n_2}{n_1}\right)\right]. \tag{7.48}$$

Similarly, the equivalent index also simplifies at the reference wavelength. Using $G=1$ in equation (7.47), we can get

$$\eta|_{G=1} = \pm jn_1. \tag{7.49}$$

We can note that $\theta|_{G=1}$ will be complex because $\frac{1}{2}\left(\frac{n_1}{n_2} + \frac{n_2}{n_1}\right)$ will always be greater than 1. Since n_1 is real, the equivalent index at $G=1$ will be imaginary. This is the same result that we stated earlier in the preview example (equation (7.5)). Although one might associate an imaginary refractive index with absorption losses, as we discussed in Chapter 1, the absorption in a film is determined only by its loss tangent value. In this case, we can verify that a layer with a purely imaginary refractive index will have a zero loss tangent (equation (1.76) in Chapter 1), and this is consistent with the expectation that the structure should be lossless because all films in the stack consist of real refractive indices.

As an example, considering $\left(\frac{H}{2}L\frac{H}{2}\right)$ with $n_1 = 2.5$ and $n_2 = 1.5$, Figures 7.5 and 7.6 show the calculated results for η and θ as a function of G. As should be evident from these figures, the equivalent index η is imaginary when $G=1$, as well as within a certain band around $G=1$. The phase thickness θ is also complex in this region.

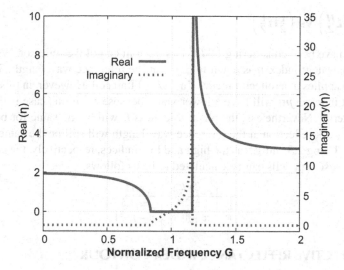

Figure 7.5 Real and imaginary parts of the equivalent index η for $\left(\frac{H}{2}L\frac{H}{2}\right)$ as a function of the normalized frequency G using $n_1 = 2.5$ and $n_2 = 1.5$.

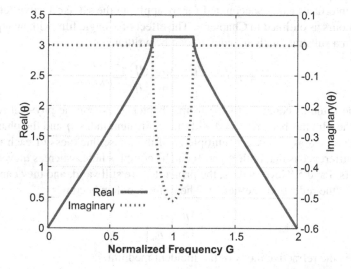

Figure 7.6 Real and imaginary parts of the equivalent phase θ for $\left(\frac{H}{2}L\frac{H}{2}\right)$ as a function of G using $n_1 = 2.5$ and $n_2 = 1.5$.

Outside of this band, both quantities become real. Due to the complex (hence exponential) behavior of the fields in this band, we define this central region as the *stop band* (or as the photonic bandgap). The imaginary part of η can be positive or negative (both are valid solutions), but we have only shown the positive values in Figure 7.5.

7.5 $\left(\frac{H}{2}L\frac{H}{2}\right)$ vs $\left(\frac{L}{2}H\frac{L}{2}\right)$

So far we have been considering $\left(\frac{H}{2}L\frac{H}{2}\right)$ as the unit cell of the reflector. We showed that its equivalent index η is equal to $\pm jn_1$ at the reference wavelength. The same result also applies to a reflector made of a $\left(\frac{L}{2}H\frac{L}{2}\right)$ unit cell. However, in this case, the equivalent index $\pm jn_1$ will have a lower value because n_1 in this case is the lower-index material. Nevertheless, the phase thickness θ will be the same for both. As a result, the peak reflection at the reference wavelength will still be the same for both structures. Using 2.5 and 1.5 as the high and low indices, respectively, the equivalent values of these unit cells can be calculated to be as follows.

	$\left(\frac{H}{2}L\frac{H}{2}\right)$	$\left(\frac{L}{2}H\frac{L}{2}\right)$
$\frac{q}{k_0}$	$\pm j2.5$	$\pm j1.5$
θ	$\pi - j0.51$	$\pi - j0.51$

7.6   EFFECTIVE REFLECTANCE INDEX CONTOUR

So far in this description, we have excluded the substrate (exit medium below the stack) and air (incident medium above the stack). We need to include these interfaces to be able to predict the reflection and transmission from the stack.

These interfaces can be accounted for by applying the same effective reflectance index contours as outlined in Chapter 3. The effect of a single film with an equivalent index η on a substrate with an index n_s can be written as

$$n_r = \eta \frac{(n_s + \eta) + (n_s - \eta)e^{-j2N\theta}}{(n_s + \eta) - (n_s - \eta)e^{-j2N\theta}}. \tag{7.50}$$

This is the same as equation (3.16) from Chapter 3, except the physical refractive index of the film has been replaced with the equivalent index η and the phase thickness is expressed as an integer multiple N of the phase thickness of each unit cell. Another difference is that both η and θ can be complex here, whereas they were real in Chapters 3 and 5. Nevertheless, the principles are still valid, and they can be used to calculate the reflection coefficient. The reflection then becomes

$$R = \left| \frac{n_r - n_a}{n_r + n_a} \right|^2, \tag{7.51}$$

where n_a is the refractive index of the incident medium.

We can also plot the effective reflectance index contour to gain some understanding similar to our earlier antireflection studies. Considering $\left(\frac{H}{2}L\frac{H}{2}\right)^N$ with $n_1 = 2.5$, $n_2 = 1.5$, and $n_s = 1.5$, the contour of n_r is shown in Figure 7.7. These are discrete points rather than a continuous line because the effective reflectance indices jump from one point to another as N increases in integer steps.

Taking a closer look at Figure 7.7, we can see that unlike the effective index contours from conventional dielectric films, this one does not make circuitous routes as the phase thickness is increased. The contour starts from the substrate index for $N = 0$ and converges toward $\eta = jn_1$ (Herpin's equivalent index) as N increases. This type

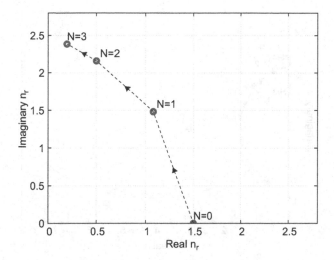

Figure 7.7 Effective reflectance index contour for $\left(\frac{H}{2}L\frac{H}{2}\right)^N$ using $n_1 = 2.5$, $n_2 = 1.5$, and $n_s = 1.5$ for increasing values of N.

of convergence is also seen in films with complex indices (similar to metals). This should not be surprising because a high-reflection dielectric stack behaves similar to a metal reflector in many respects. This situation will be encountered again later in the context of metal films in Chapter 12.

Since n_r converges toward jn_1 (and is independent of n_2), one should not conclude that the low-index film n_2 has no effect on this contour. Both n_1 and n_2 are included in the expression for the phase thickness θ, which is the distance between each discrete step in Figure 7.7. In other words, the difference between n_1 and n_2 will dictate how quickly n_r approaches its final value of jn_1. This should make sense because a unit cell with a high index contrast will approach its high reflective condition faster than a unit cell with a lower contrast.

It is also instructive to examine the data in Figure 7.7 along with the contour plot of treating each layer separately as we did in Chapter 6. This is shown in Figure 7.8. Both contours start at $n_r = n_s$ but have different trajectories. However, both contours have the same end points at the end of each unit cell. In other words, the Herpin's method gives us a convenient way to trace the end points of each unit cell without having to make multiple circuitous contours. In addition, the fact that all unit cells have the same equivalent index η is a significant advantage when we design structures such as edge filters, as we will see in Chapter 8.

7.7   REFLECTION AND TRANSMISSION AT THE REFERENCE WAVELENGTH

At the reference wavelength, we can simplify the expression for n_r if we assume the substrate index n_s is the same as n_1. Since $\eta = jn_1$, we can use this to significantly simplify equation (7.50), resulting in

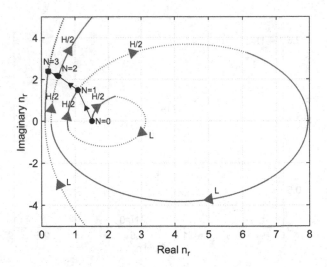

Figure 7.8 Effective reflectance index contour for $\left(\frac{H}{2}L\frac{H}{2}\right)^N$ treating each layer separately and comparing it to the Herpin's model.

$$n_r = n_1 \left(\frac{\cos\theta - \sin\theta}{\cos\theta + \sin\theta} \right). \tag{7.52}$$

Substituting this into equation (7.51), we can get the expression for the reflection at the reference wavelength:

$$R = \left| \frac{1 - \left(\frac{\cos\theta - \sin\theta}{\cos\theta + \sin\theta}\right)}{1 + \left(\frac{\cos\theta - \sin\theta}{\cos\theta + \sin\theta}\right)} \right|^2 \tag{7.53}$$

$$= \left| \frac{\sin\theta}{\cos\theta} \right|^2 \tag{7.54}$$

$$= |\tan\theta|^2. \tag{7.55}$$

If there are N unit cells in the stack, the reflection becomes simply

$$R = |\tan N\theta|^2. \tag{7.56}$$

This is an extremely simple expression, with the caveat that the incident medium and the substrate are assumed to have the same refractive index as n_1. For a different substrate index and incident medium, equation (7.50) has to be evaluated. However, for a high reflection stack, the incident and exit mediums have minimal impact on the overall reflectance, which can be easily verified.

7.7.1 STOP BAND

The stop band (or photonic bandgap) is the spectral region where the fields have an exponential character. This is the region in Figures 7.5 and 7.6 where θ is complex and η is imaginary. We can see from these figures that θ becomes π at both edges of the stop band, and η becomes zero at one edge of the stop band, and ∞ at the other end. Therefore, we can solve for the edges of the stop band by setting $\theta = \pi$ in equation (7.46):

$$\cos(\pi) = 1 - \frac{(n_1 + n_2)^2}{2n_1 n_2} \sin^2\left(\frac{G\pi}{2}\right) \tag{7.57}$$

$$G = \left(\frac{2}{\pi}\right) \sin^{-1}\left(\frac{2\sqrt{n_1 n_2}}{n_1 + n_2}\right). \tag{7.58}$$

This has two solutions on either side of $G = 1$, corresponding to

$$G = \left(\frac{2}{\pi}\right) \sin^{-1}\left(\frac{2\sqrt{n_1 n_2}}{n_1 + n_2}\right) \tag{7.59}$$

$$G = \left(\frac{2}{\pi}\right)\left[\pi - \sin^{-1}\left(\frac{2\sqrt{n_1 n_2}}{n_1 + n_2}\right)\right]. \tag{7.60}$$

The spectral width of the stop band is the difference between the two solutions:

$$\Delta G = \left(\frac{2}{\pi}\right)\left[\pi - 2\sin^{-1}\left(\frac{2\sqrt{n_1 n_2}}{n_1 + n_2}\right)\right]. \tag{7.61}$$

We can also write n_2 as $n_1 + \Delta n$ where Δn is the refractive index contrast between the two layers. We can then show that

$$\sin^{-1}\left(\frac{2\sqrt{n_1 n_2}}{n_1 + n_2}\right) = \sin^{-1}\left(\frac{\sqrt{1 + \Delta n/n_1}}{1 + \Delta n/2n_1}\right). \tag{7.62}$$

From this, we can see that if Δn is equal to zero, then the argument of $\sin^{-1}$ will become 1, which makes ΔG equal to zero. This is consistent with our understanding that the reflection bandwidth will collapse to zero when the refractive index contrast disappears. At the other extreme, the maximum value of G will occur when the $\sin^{-1}$ is equal to 0 or when Δn is infinitely large. This will produce the maximum possible value of G,

$$\Delta G|_{\text{max}} = 2.0, \tag{7.63}$$

which spans the frequency range of $G = 0$ to $G = 2$.

In terms of wavelength, the band edges can be written as

$$\lambda = \lambda_0 \left[\frac{\pi/2}{\sin^{-1}\left(\frac{\sqrt{1 + \Delta n/n_1}}{1 + \Delta n/2n_1}\right)}\right] \tag{7.64}$$

$$\lambda = \lambda_0 \left[\frac{\pi/2}{\pi - \sin^{-1}\left(\frac{\sqrt{1 + \Delta n/n_1}}{1 + \Delta n/2n_1}\right)}\right] \tag{7.65}$$

or

$$\Delta\lambda = \lambda_0 \left(\frac{\pi}{2}\right) \left[\frac{1}{\sin^{-1}\left(\frac{\sqrt{1+\Delta n/n_1}}{1+\Delta n/2n_1}\right)} - \frac{1}{\pi - \sin^{-1}\left(\frac{\sqrt{1+\Delta n/n_1}}{1+\Delta n/2n_1}\right)} \right]. \tag{7.66}$$

While $\Delta G|_{\max}$ is limited to 2.0, the maximum value of $\Delta\lambda$ is unlimited. This is because of the inverse relationship between frequency and wavelength. While $G = 2$ corresponds to half the reference wavelength, $G = 0$ corresponds to an infinitely large wavelength. Therefore, if we were able to have an infinitely large refractive index contrast between the refractive indices of the high- and low-index films, the reflection bandwidth will extend from $\lambda_0/2$ on the left side to indefinite long-wavelength values on the right hand side. This is illustrated in Figure 7.9. We can see that G moves symmetrically about the normalized frequency as the refractive index contrast is increased but λ moves more rapidly on the long-wavelength side compared to the short-wavelength side.

From Figure 7.5, we can see that the equivalent index is zero at the left edge of the stop band and infinitely large at the right edge. We will later see that when the unit cell is $\left(\frac{L}{2}H\frac{L}{2}\right)$, these positions of zero and infinite η will be reversed. This occurs because the sign of $(n_1 - n_2)$ in equation (7.45) will reverse, resulting in the numerator and denominator exchanging positions. The zero and infinite equivalent indices are not a cause for concern because the phase thicknesses are still finite (and equal to π). This absentee condition combined with the very large or very small equivalent index results in large reflection coefficients at the ends of the structure. This results in a resonant condition, which can be utilized in lasers. In fact, this is the basis of distributed feedback (DFB) lasers.

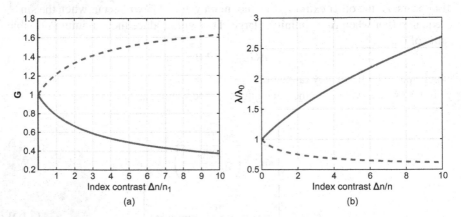

Figure 7.9 (a) Normalized frequencies and (b) their corresponding wavelengths of the band edges as a function of refractive index contrast.

7.8   REFLECTION AT THE EDGES OF THE STOP BAND

We can calculate the values of n_r (and subsequently the reflection) at the edges of the stop band. Interestingly, a phase thickness of π corresponds to the absentee condition. That means, at the edges of the stop band, the unit cell behaves as an absentee layer. It may seem that the equivalent index layer should simply disappear when $\theta = \pi$, simplifying the expression for n_r. However, since η asymptotically approaches zero and ∞ when $\theta \to \pi$, this calculation is not that straightforward. It requires us to take the limiting values.

As $\theta \to \pi$, equation (7.50) can be written as

$$n_r|_{\theta \to \pi} = \eta \frac{(n_s + \eta) + (n_s - \eta)e^{-j2N\theta}}{(n_s + \eta) - (n_s - \eta)e^{-j2N\theta}}\Bigg|_{\theta \to \pi} \tag{7.67}$$

$$= \frac{2n_s + \eta\left(1 - e^{-j2N\theta}\right)}{2 + \frac{n_s}{\eta}\left(1 - e^{-j2N\theta}\right)}\Bigg|_{\theta \to \pi}. \tag{7.68}$$

Taking the case where $\eta - \to 0$ as $\theta \to \pi$,

$$n_r|_{\theta \to \pi, \eta \to 0} = \frac{2n_s}{2 + n_s\frac{(1 - e^{-j2N\theta})}{\eta}}\Bigg|_{\theta \to \pi}. \tag{7.69}$$

The limiting value of $\frac{(1 - e^{-j2N\theta})}{\eta}$ has to be calculated because both the numerator and denominator approach zero for $\theta \to \pi$. For this, we can write θ as $\pi + \Delta$ where $\Delta \to 0$. Then using Taylor's expansion of cosine for a small angle Δ in the expression for η, and then applying L'Hospital's rule, we can get

$$\frac{\left(1 - e^{-j2N\Delta}\right)}{\eta}\Bigg|_{\Delta \to 0} = j4N\frac{(1 - n_2/n_1)}{\sqrt{n_1 n_2}}. \tag{7.70}$$

This results in

$$n_r|_{\theta \to \pi, \eta \to 0} = \frac{n_s}{1 + j2n_s N\frac{(1 - n_2/n_1)}{\sqrt{n_1 n_2}}}. \tag{7.71}$$

Using a similar approach for the case where $\eta - \to \infty$ as $\theta \to \pi$, we can get

$$n_r|_{\theta \to \pi, \eta \to \infty} = n_s + \frac{\left(1 - e^{-j2N\theta}\right)}{2(1/\eta)}\Bigg|_{\theta \to \pi}. \tag{7.72}$$

This results in

$$n_r|_{\theta \to \pi, \eta \to \infty} = n_s + j2N\frac{(n_1 - n_2)}{\sqrt{n_1 n_2}}. \tag{7.73}$$

Now we can apply both these values of n_r into equation (7.51) and calculate the reflection values at the edges of the stop band.

Additionally, we can see that $n_r|_{\theta \to \pi, \eta \to 0}$ approaches 0 as the number of unit cells N approaches a large number. Similarly, $n_r|_{\theta \to \pi, \eta \to \infty}$ approaches ∞ for large N. Both of these cases would produce a reflection approaching unity. In other words, we can state that the reflection spectrum within the stop band will approach a flat top profile with values close to 100% as $N \to \infty$.

7.8.1 HIGHER-ORDER ABSENTEE CONDITIONS

The stop band was defined in terms of each unit cell becoming an absentee layer ($\theta = \pi$). However, that is only one of several possible absentee conditions when the structure consists of multiple unit cells. Considering that there are N unit cells in the stack, the total phase thickness will be $N\theta$. As long as $N\theta$ is an integer multiple of π, it should satisfy the absentee condition. We can write this as

$$m\pi = N\theta \tag{7.74}$$

$$= N\cos^{-1}\left[1 - \frac{(n_1 + n_2)^2}{2n_1 n_2}\sin^2\left(\frac{G\pi}{2}\right)\right], \tag{7.75}$$

where m is a positive integer. The stop band condition of $\theta = \pi$ corresponds to $m = N$, so that the N cancels out on both sides of the equation.

Clearly, there are other values of m that also satisfy the absentee condition. Furthermore, the equivalent index $\frac{q}{k_0}$ has asymptotically approaching values only for $m = N$. For all other $m \neq N$, $\frac{q}{k_0}$ is a well-behaved function with finite values (see Figure 7.5). As a result, n_r would become equal to n_s for all $m \neq N$ producing a reflection minima at these points. The pair of frequencies that correspond to this absentee condition can be written as

$$G = \frac{2}{\pi}\sin^{-1}\left\{\frac{\sqrt{2n_1 n_2\left(1 - \cos\left(\frac{m\pi}{N}\right)\right)}}{n_1 + n_2}\right\} \tag{7.76}$$

for the lower frequency and

$$G = \frac{2}{\pi}\left[\pi - \sin^{-1}\left\{\frac{\sqrt{2n_1 n_2\left(1 - \cos\left(\frac{m\pi}{N}\right)\right)}}{n_1 + n_2}\right\}\right] \tag{7.77}$$

for the higher frequency.

In terms of Δn, this can be written as

$$G = \frac{2}{\pi}\sin^{-1}\left\{\frac{\sqrt{(1 + \Delta n/n_1)\frac{(1 - \cos(\frac{m\pi}{N}))}{2}}}{1 + \Delta n/2n_1}\right\} \tag{7.78}$$

$$G = \frac{2}{\pi}\left[\pi - \sin^{-1}\left\{\frac{\sqrt{(1 + \Delta n/n_1)\frac{(1 - \cos(\frac{m\pi}{N}))}{2}}}{1 + \Delta n/2n_1}\right\}\right]. \tag{7.79}$$

7.9 EXAMPLE – CONTINUED FROM SECTION 7.2

Let's reconsider the example from Section 7.2, which was a periodic stack $\left(\frac{H}{2}L\frac{H}{2}\right)^5$ with $n_1 = 2.5$, $n_2 = 1.5$, and $n_s = 1.5$.

We can use equation (7.75) to calculate all of the absentee values. This is shown in Table 7.1. The values for $m = 5$ correspond to the previously calculated stop band edges, $m = 4$ corresponds to the first minima outside the reflection band, $m = 3$ corresponds to the second minima, and so on.

Table 7.1

Calculated Values of Normalized Frequency G and Their Corresponding Wavelengths for the Different Absentee Layer Conditions Using $\left(\frac{H}{2}L\frac{H}{2}\right)^5$ with $n_1 = 2.5$ and $n_2 = 1.5$. These Are Shown on Figure 7.10

	m	G	λ (nm)	G	λ (nm)
Stop band	5	0.8391	655.4	1.1609	473.8
First min.	4	0.7450	738.2	1.2550	438.2
Second min.	3	0.5730	959.9	1.4270	385.4
Third min.	2	0.3854	1427	1.6146	340.6
Fourth min.	1	0.1934	2843	1.8066	304.4

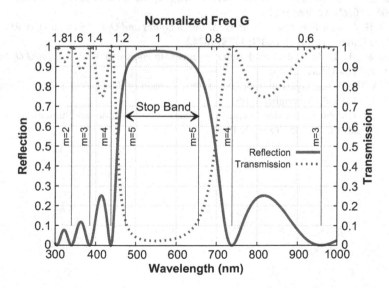

Figure 7.10 Reflection and transmission spectra for $\left(\frac{H}{2}L\frac{H}{2}\right)^5$ with $n_1 = 2.5$, $n_2 = 1.5$, and $n_s = n_a = 2.5$.

7.10 PROBLEMS

1. Plot the equivalent index and phase thickness of a three-layer structure consisting of $Si–SiO_2–Si$, as well as $SiO_2–Si–SiO_2$, at a reference wavelength of 1.55 μm.

2. For a periodic stack that consists of five repeating unit cells of the above two structures, calculate the peak reflection at the reference wavelength, the width of the stop band, and the first points of zero reflectance outside the stop band.

3. Consider a periodic structure consisting of MgF_2-Al_2O_3-MgF_2 for a reference wavelength of 800 nm. Assuming the substrate index is 1.48, plot the reflection spectra for values of N from 1 to 20. Calculate the reflection values at the edges of the stop band using TMM, and verify that they match with the results from equations (7.71) and (7.73). Also calculate the peak reflection at $G = 1$, and verify that it matches with that of the TMM.

4. An application requires a stop band width of ~100 nm with a reference wavelength of 750 nm. If one of the layer materials is SiO_2, find which material can serve as the second material.

5. A resonance condition in an optical structure can be thought of as when the total phase thickness is $N\pi$, combined with a relatively long photon life time. Show that both conditions are satisfied at the two edges of the stop band. These are the cavity resonance wavelengths of a DFB laser.

FURTHER READING

Epstein, L. I. The design of optical filters. *Journal of the Optical Society of America* **42**, 806–810, (November 1952).

Macleod, H. A. *Thin-Film Optical Filters (Series in Optics and Optoelectronics)*, (CRC Press, Boca Raton, FL, 2017). ISBN: 1138198242.

Schallenberg, U. B. Antireflection design concepts with equivalent layers. *Applied Optics* **45**, 1507–1514, (March 2006).

Thelen, A. Equivalent layers in multilayer filters. *Journal of the Optical Society of America* **56**, 1533–1538, (November 1966).

Willey, R. R. Graphic description of equivalent index approximations and limitations. *Applied Optics* **28**, 4432–4435, (October 1989).

8 Edge Filters

8.1 BASIC CONCEPTS

Edge filters can be viewed as consisting of a high-reflection band adjacent to an antireflection band. These are also known as dichroic filters or as hot and cold mirrors in the context of infrared radiation (Figure 8.1). As discussed in Chapters 6 and 7, the performance of the high reflector is determined primarily by the refractive index contrast between the high-index and low-index layers and the number of repeating periods used in the design. From Figure 6.4 in Chapter 6, we also noted that one side of the passband exhibited strong oscillations and the other side exhibited a flatter reflection spectrum. The Herpin equivalent index discussed in Chapter 7 provides useful insights into the origins of these oscillations and means of suppressing them.

8.2 EQUIVALENT INDEX OF THE PASSBAND OF A PERIODIC STACK

The majority of the periodic thin film calculations are performed for the reflection band (or stop band) centered around the reference wavelength, $G = 1$. In the case of edge filters, the properties of the passband just outside the stop band are just as important. However, this region is not simple to quantify mathematically because it does not have a specific relationship to the reference wavelength. As a result,

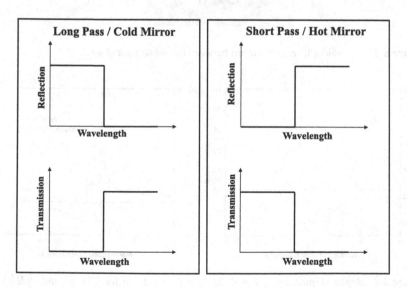

Figure 8.1 Long-pass and short-pass filters (also known as cold and hot mirrors).

the effective reflectance index discussed in Chapters 3–6 is not useful here. But the Herpin equivalent index comes in handy for this purpose.

Let's consider a simple model where a film with an equivalent index $\eta = \frac{q}{k_0}$ and equivalent phase thickness $N\theta$ is sandwiched between the two outside media (n_a, η, and n_s). As shown in Figure 8.2, this can be viewed as a simple Fabry–Pérot cavity. The strength of the resonant oscillations will depend on the refractive index difference between the interfaces, and the spacing of the oscillation peaks will depend on $N\theta$. While it is possible to achieve high transmission at the cavity resonant frequencies, in order to achieve a flat transmission profile over a broad spectral range, all three indices (or their effective values) must become close to each other.

For $\left(\frac{H}{2}L\frac{H}{2}\right)$, the Herpin equivalent index η was ∞ at the low-frequency (long-wavelength) edge of the reflection band and 0 at the higher-frequency (short-wavelength) edge. Using $n_1 = 2.5$ and $n_2 = 1.5$, these edges were calculated as $G = 0.8391$ and $G = 1.1609$. On Figure 8.3, the equivalent index as a function of G for both $\left(\frac{H}{2}L\frac{H}{2}\right)$ and $\left(\frac{L}{2}H\frac{L}{2}\right)$ are shown.

As discussed in Chapter 7, the stop band is characterized by a purely imaginary equivalent index η. Outside this stop band, the equivalent index becomes real. For $\left(\frac{H}{2}L\frac{H}{2}\right)$, the left side has a smaller average equivalent index than the right side.

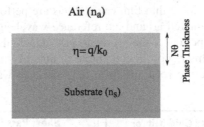

Air (n_a)

$\eta = q/k_0$

$N\theta$
Phase Thickness

Substrate (n_s)

Figure 8.2 Equivalent film sandwiched between the substrate and air.

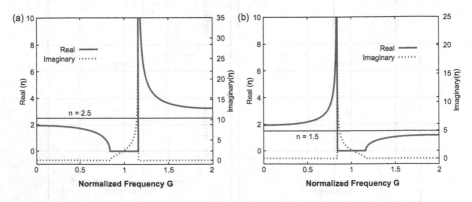

Figure 8.3 Real and imaginary parts of the equivalent index η for $\left(\frac{H}{2}L\frac{H}{2}\right)$ and $\left(\frac{L}{2}H\frac{L}{2}\right)$ unit cells using 2.5 and 1.5 for the high- and low-index materials. (a) $\left(\frac{H}{2}L\frac{H}{2}\right)$. (b) $\left(\frac{L}{2}H\frac{L}{2}\right)$.

The opposite is true for $\left(\frac{L}{2}H\frac{L}{2}\right)$. These are in reference to the normalized frequency G, so the left and right will be reversed when plotted against wavelength. Also note that Figure 8.3a and b are not exactly a mirror image of each other. While the long-wavelength side of $\left(\frac{H}{2}L\frac{H}{2}\right)$ spans from an index value of 0 to 2, the short-wavelength side of $\left(\frac{L}{2}H\frac{L}{2}\right)$ spans from 0 to only about 1. This aspect will result in a slightly different passband behavior between the two structures.

We can also make the observation that the equivalent indices on either side of the stop band straddle the refractive index value of the outer layer of the unit cell. This is shown by the dotted line in Figure 8.3a and b. For example, for $\left(\frac{H}{2}L\frac{H}{2}\right)$, the index on the left is about 2.0 (lower than the H layer index of 2.5), and 4.0 on the right (higher than 2.5). Similarly, for $\left(\frac{L}{2}H\frac{L}{2}\right)$, the indices are about 2.0 (higher than 1.5) and 1.2 (lower than 1.5). These observations will provide some guidelines when designing matching layers to reduce feedback.

We can also plot the phase thickness θ as a function of G. This is shown in Figure 8.4 (same as Figure 7.6 in Chapter 7). This plot is identical for both $\left(\frac{L}{2}H\frac{L}{2}\right)$ and $\left(\frac{H}{2}L\frac{H}{2}\right)$, so only one is shown. We can see that the phase thickness is real-valued outside the reflection band and is imaginary-valued inside (since the real part is equal to π). Furthermore, we can also note that the phase thickness is symmetric about $G = 1$.

When designing a long-pass filter, the left side of the plots in Figure 8.3 would be the relevant region. The index values here are from 0 to 2.0 for $\left(\frac{H}{2}L\frac{H}{2}\right)$ and from $+\infty$ to about 2.0 for $\left(\frac{L}{2}H\frac{L}{2}\right)$. If the substrate index is 1.5, $\left(\frac{H}{2}L\frac{H}{2}\right)$ would have a better match to the substrate index on the long-wavelength side of the stop band. $\left(\frac{L}{2}H\frac{L}{2}\right)$ will have a worse match.

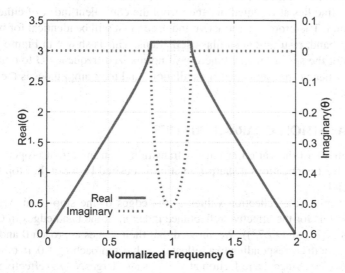

Figure 8.4 Real and imaginary parts of the equivalent phase thickness, using 2.5 and 1.5 for the high- and low-index materials.

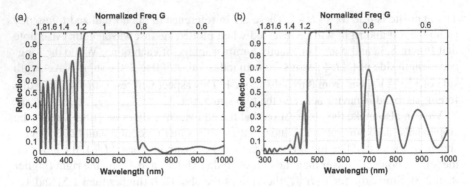

Figure 8.5 Reflection spectra of $\left(\frac{H}{2}L\frac{H}{2}\right)^{10}$ and $\left(\frac{L}{2}H\frac{L}{2}\right)^{10}$ with film indices of 2.5 and 1.5 on a substrate with index 1.5. (a) $\left(\frac{H}{2}L\frac{H}{2}\right)^{10}$. (b) $\left(\frac{L}{2}H\frac{L}{2}\right)^{10}$.

To illustrate this, Figure 8.5a shows $\left(\frac{H}{2}L\frac{H}{2}\right)^{10}$ using $n_1 = 2.5$ and $n_2 = 1.5$ on a substrate with $n_s = 1.5$. Figure 8.5b shows $\left(\frac{L}{2}H\frac{L}{2}\right)^{10}$ on the same substrate. It is clear from Figure 8.5a that $\left(\frac{H}{2}L\frac{H}{2}\right)^{10}$ is suitable as a long-pass filter. Additionally, Figure 8.5b shows that although it can be used as a short-pass filter, it does have some prominent ripples in the passband. This is because the equivalent index of $\left(\frac{L}{2}H\frac{L}{2}\right)$ on the short-wavelength side spans from 0 to 1 and has a worse match with the substrate than the long-wavelength side of $\left(\frac{H}{2}L\frac{H}{2}\right)$.

Interestingly, if the substrate index and incident medium have the same index as the first film in the trilayer sequence, i.e., $n_s = n_a = n_1$, then the resulting reflection spectrum will be symmetric. The reason for this should be apparent from Figure 8.3a. The dotted line lies at an equal distance from the equivalent index of either side of the stop band. Therefore, the refractive index contrast will be identical for both sides of the stop band, resulting in similar oscillations. This is shown in Figure 8.6. This time we plot the spectrum as a function of normalized frequency G to demonstrate that the frequency spacing between oscillations and their amplitudes is the same on both sides.

8.3   TRANSITION CHARACTERISTICS

The abruptness of the cut-on or cut-off transition is an important property of edge filters. In most applications a sharp transition is desired between the stop band and the passband.

Let's examine the reflection values at the edges of the stop band. We derived the expressions for the effective reflectance index n_r at the band edges in Chapter 7 (equations (7.71) and (7.73)). We showed that their values approach 0 and ∞ at the band edges, both corresponding to a reflection value approaching 1.0. In other words, $R \to 1$ as $N \to \infty$. Since the reflection at $G = 1$ is also large, $N \to \infty$ effectively makes the reflection spectrum of the stop band flatter, producing a rectangular shape. On the other hand, the next absentee condition ($m = N - 1$) is a minimum reflection point.

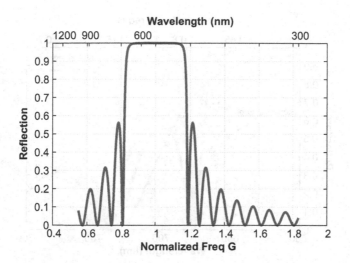

Figure 8.6 Reflection spectrum of $\left(\frac{H}{2}L\frac{H}{2}\right)^{10}$ with $n_1 = 2.5$, $n_2 = 1.5$, $n_s = n_a = 2.5$, and $\lambda_0 = 550$ nm.

Therefore, we can use the spectral distance between the stop band edge $(m = N)$ and the first minimum $(m = N - 1)$ as a measure of the transition gradient.

Using equations (7.76) and (7.77) in Chapter 7, we can plot the normalized frequencies of the edge of the stop band and the first minimum outside the stop band. This is shown in Figure 8.7 for $n_1 = 2.5$ and $n_2 = 1.5$. We can see that the edge of the stop band remains constant while the first minimum approaches closer toward the

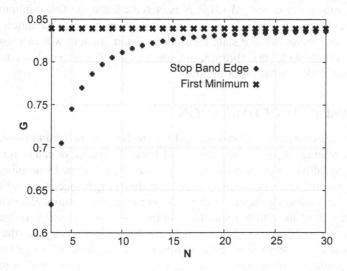

Figure 8.7 Normalized frequency G of the stop band edge $(M = N)$ and the first minimum $(M = N - 1)$ as a function of the number of unit cells, assuming $n_1 = 2.5$ and $n_2 = 1.5$.

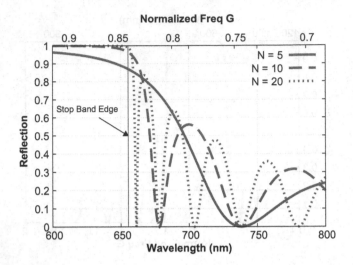

Figure 8.8 Long-wavelength transition edge of $\left(\frac{H}{2}L\frac{H}{2}\right)^N$ as a function of N using $n_1 = 2.5$, $n_2 = 1.5$, $n_a = n_s = 2.5$, and $\lambda_0 = 550$ nm.

stop band edge as N increases. This rate of closure depends on the refractive index difference between n_1 and n_2. The difference between the two G values will basically define the steepness of the transition at the band edge.

Figure 8.8 shows the long-wavelength transition edge of $\left(\frac{H}{2}L\frac{H}{2}\right)^N$ for different values of N, using $n_1 = 2.5$, $n_2 = 1.5$ with $n_a = n_s = 2.5$ (to maintain symmetry between the two transition edges). We can see that the reflection at the stop band edge moves closer toward 1.0 as N is increased, and the first minimum moves closer toward the stop band edge, which effectively steepens the transition gradient. Therefore, in general, we can state that the transition gradient will increase as N is increased. Filters with very abrupt transition edges utilize a very large number of repeating unit cells, on the order of 50.

8.4   NUMERICAL OPTIMIZATION

The designs discussed in Sections 8.2 and 8.3 can be used as starting points for creating more refined long-pass and short-pass filters. These designs can be numerically optimized by adjusting every layer in the structure. This is done by defining an error function – the difference between a desired spectrum (target spectrum) and the actual spectrum – and allowing the numerical process to execute a minimization algorithm by adjusting all of the allowed variables. In this case, the allowed variables are the film thicknesses. The refractive indices can also be allowed to vary, but that is a less practical solution unless we have the ability to tune the material index (which can be done by mixing different materials). A large number of numerical minimization routines are available, the details of which are outside the scope of this book [1–4]. The example code given in Chapter 15 shows a demonstration of this optimization

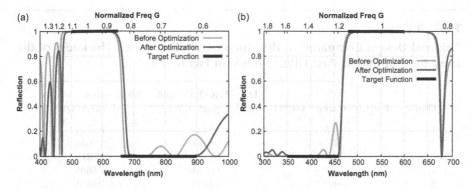

Figure 8.9 Numerically optimized $\left(\frac{H}{2}L\frac{H}{2}\right)^{10}$ long-pass and $\left(\frac{L}{2}H\frac{L}{2}\right)^{10}$ short-pass filters on a substrate with index of 1.0. (a) $\left(\frac{H}{2}L\frac{H}{2}\right)^{10}$ long-pass filter. (b) $\left(\frac{L}{2}H\frac{L}{2}\right)^{10}$ short-pass filter.

technique. Additionally, the freely available OpenFilters software [5] also includes much more efficient optimization routines.

Figure 8.9a and b shows the numerically refined long-pass and short-pass filters using $\left(\frac{H}{2}L\frac{H}{2}\right)^{10}$ and $\left(\frac{L}{2}H\frac{L}{2}\right)^{10}$ structures as the baseline design with high-index and low-index materials having indices of 2.5 and 1.5, respectively, and a substrate with an effective index of $n_s = 1.0$. The target reflection functions and the optimized results are also shown on the same plot. The optimization is performed by adjusting each of the 21 layers in the design while keeping their refractive indices constant. After optimization, none of the layers will be quarter wave thick. However, most of the layers of the optimized structure will be within a small range of the original film thicknesses, which gives validation to our starting design. The usable range of the long-pass filter is from 500 to 900 nm, with a cut-on transition at 650–675 nm. For the short-pass filter, the usable range is from 350 to 600 nm, with the cut-off transition at 460–480 nm. The starting and ending film thicknesses for all 21 layers after optimization are shown in Table 8.1.

8.5 EFFECTS OF MATERIAL DISPERSION

Thus far, we have assumed the films to have a fixed refractive index for all wavelengths. Although this is useful for the initial evaluation of the design, real materials have dispersion, which will inevitably affect the performance of the filter. However, just as the undesired ripples in the passband were removed by numerical optimization, minor dispersion effects can also be easily corrected with optimization.

For example, Figure 8.10 shows the $\left(\frac{L}{2}H\frac{L}{2}\right)^{10}$ short-pass filter from Figure 8.9b, but this time using dispersive refractive indices of TiO_2 and SiO_2 for the high- and low-index films, with a SiO_2 substrate. The baseline and optimized performances are shown. Since both materials begin to exhibit significant dispersion at wavelengths closer to 400 nm, optimization will become particularly important.

Table 8.1

Original Design Compared to the Numerically Optimized Thickness of the Long-Pass and Short-Pass Filters Shown in Figure 8.9

Film #	Original Thickness (QW)	Long-Pass Optimized Thickness (QW)	Short-Pass Optimized Thickness (QW)
1	0.5	0.3428	0.5392
2	1	1.0598	1.0634
3	1	1.1716	1.0211
4	1	1.0753	1.0806
5	1	0.8884	0.9843
6	1	0.8802	0.9991
7	1	1.0817	1.0229
8	1	1.0539	0.9852
9	1	1.0801	0.9786
10	1	0.9937	1.0165
11	1	0.9355	0.9664
12	1	0.9887	1.0232
13	1	1.0768	0.9556
14	1	1.0688	1.0196
15	1	1.0757	1.0045
16	1	0.8677	0.9803
17	1	0.9095	1.0340
18	1	1.0519	1.0233
19	1	1.2041	1.0665
20	1	1.0327	1.0401
21	0.5	0.3540	0.5448

8.6   DESIGN EXAMPLE OF A MID-INFRARED LONG-PASS EDGE FILTER

The design principles remain the same regardless of the spectral range. The materials, however, will change. For example, consider a cut-on filter on a ZnS substrate for the mid-infrared spectrum (3–5 μm), with a transition at 4 μm, with the additional requirements of >99% reflection in the stop band and >99% transmission in the passband with a transition width <100 nm.

The refractive index of ZnS at 4 μm is 2.25. The first step in the design process is to identify two suitable thin-film materials that can serve as the high-index and low-index materials. Ge and MgF_2 can be considered, since they are both transparent in the 3–5 μm spectral range and can be deposited relatively easily. Most importantly, their refractive indices are 4.0 and 1.32, respectively. The large difference between their refractive indices should allow sufficiently high reflectivity to be achieved with the fewest number of layers.

From equations (7.48) and (7.56) in Chapter 7, assuming the substrate index is the same as the outer layers of the unit cell, we can estimate the peak reflection at the reference wavelength as

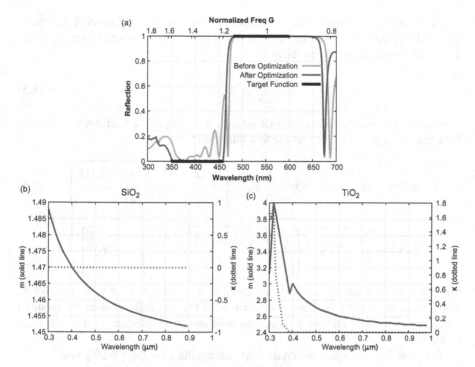

Figure 8.10 Optimized short-pass filter $\left(\frac{L}{2}H\frac{L}{2}\right)^{10}$ using SiO_2 and TiO_2 with dispersion.

$$R = \left| \tan\left(N\cos^{-1}\left\{ -\frac{1}{2}\left(\frac{n_1}{n_2} + \frac{n_2}{n_1} \right) \right\} \right) \right|^2. \tag{8.1}$$

We can solve for integer values of N for achieving $R > 0.99$. For $N = 3$, the reflection works out to $R = 0.995$. Therefore, only three unit cells are needed to achieve this peak reflection. However, as we will see, it turns out that $N = 3$ is not quite adequate to achieve the specified transition gradient.

The next step is to determine the reference wavelength. Obviously, this is not the same as the required cut-on wavelength of 4 μm. The reference wavelength will be shorter than 4 μm. Using equation (7.76) from Chapter 7, we derive the expression for the long-wavelength (low frequency) edge of the stop band:

$$G = \frac{2}{\pi}\sin^{-1}\left\{ \frac{2\sqrt{n_1 n_2}}{n_1 + n_2} \right\}. \tag{8.2}$$

Using $n_1 = 4.0$, $n_2 = 1.32$, we can get $G = 0.6639$. Since we know this edge is at a wavelength of 4,000 nm, we can use $\frac{\lambda_0}{4,000} = 0.6639$, which results in $\lambda_0 = 2,655$ nm. Although this reference wavelength falls well outside the desired operational band of 3–4 μm, it is not a cause for concern. The purpose of this reference wavelength is for defining the film thicknesses only.

The next step is to determine the number of unit cells required to meet the transition width of 100 nm. Assuming that the transition width $\Delta\lambda$ is small, we can relate it to the normalized frequency using

$$\frac{\Delta G}{\Delta\lambda} = -\frac{\lambda_0}{\lambda^2}. \tag{8.3}$$

Again, using equation (7.76) from Chapter 7, we can get the frequency difference between the stop-band edge and the first minimum:

$$\Delta G = -\Delta\lambda \frac{\lambda^2}{\lambda_0} = 0.6639 - \frac{2}{\pi}\sin^{-1}\left\{\frac{\sqrt{2n_1 n_2 \left(1 - \cos\left(\frac{N-1}{N}\pi\right)\right)}}{n_1 + n_2}\right\} \tag{8.4}$$

$$\Delta\lambda = \frac{\lambda_0}{\lambda^2}\left[\frac{2}{\pi}\sin^{-1}\left\{\frac{\sqrt{2n_1 n_2 \left(1 - \cos\left(\frac{N-1}{N}\pi\right)\right)}}{n_1 + n_2}\right\} - 0.6639\right]. \tag{8.5}$$

Since $\Delta\lambda = -100$ nm, $\lambda_0 = 2,655$ nm, and $\lambda = 4,000$ nm, we can now solve for N. This results in $N = 10$. In other words, we need ten unit cells to meet the required transition gradient, even though three unit cells would be sufficient to meet the peak reflectivity.

The next step is to plot the equivalent index values for the $\left(\frac{H}{2}L\frac{H}{2}\right)$ and $\left(\frac{L}{2}H\frac{L}{2}\right)$ unit cells to determine which structure is most appropriate for the long-pass filter on the ZnS substrate. These are shown in Figure 8.11. For the long-pass filter, we are only interested in the left side of the stop band. The left side of $\left(\frac{H}{2}L\frac{H}{2}\right)$ ranges from 0 (at the edge of the stop band) to 2.2 (at very long wavelengths), and the left side of $\left(\frac{L}{2}H\frac{L}{2}\right)$ ranges from ∞ (at the stop-band edge) to 2.2 (at very long wavelengths). The approximate average index for $\left(\frac{H}{2}L\frac{H}{2}\right)$ in the passband region of

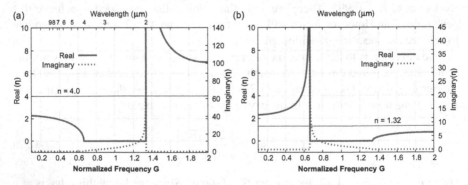

Figure 8.11 Equivalent index η for $\left(\frac{H}{2}L\frac{H}{2}\right)$ and $\left(\frac{L}{2}H\frac{L}{2}\right)$ using Ge and MgF$_2$ as the thin-film materials (refractive indices of 4.0 and 1.32, respectively, at 4,000 nm). (a) $\left(\frac{H}{2}L\frac{H}{2}\right)$. (b) $\left(\frac{L}{2}H\frac{L}{2}\right)$.

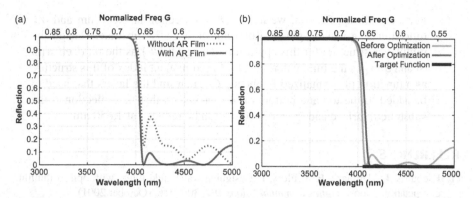

Figure 8.12 Design of a long-pass filter for the mid-infrared wavelength with a transition at $\lambda = 4$ μm. (a) $\left(\frac{H}{2}L\frac{H}{2}\right)^{10}$ with $n_1 = 4.0$, $n_2 = 1.32$, and $n_s = 2.25$. (b) Optimized with the antireflection film.

interest is about 1.0, and for $\left(\frac{L}{2}H\frac{L}{2}\right)$, it is much higher, about 8. Although neither side is ideally suited for the substrate index of 2.25, $\left(\frac{H}{2}L\frac{H}{2}\right)$ is closer to the desired 2.25. Now, referring to Figure 8.2, the average passband index of 1.0 should ideally be matched to the top interface with air and the bottom interface with the substrate. The top interface is already matched, because n_a is equal to 1.0, but the bottom interface is not. However, we can match the bottom interface by inserting an extra layer on the substrate to raise the effective reflectance index of the substrate to 2.25. This could be done with a single quarter-wave layer of index $\sqrt{2.25 \times 1.0} = 1.5$. This could be SiO$_2$ with a thickness of $\frac{4,000}{4 \times 1.5} = 666.7$ nm. Notice that the reference wavelength used here is not the same wavelength used for defining the high-reflection stack.

Figure 8.12a shows the performance of this filter with and without the 667 nm antireflection layer at the bottom. We can clearly see that the spectrum just outside the stop band is much flatter with the antireflection film. Figure 8.12b shows the performance after numerical optimization. As this example demonstrates, we can achieve a very good long-pass performance with just 22 layers of MgF$_2$ and Ge.

8.7 PROBLEMS

1. Design a long-pass edge filter with a transition wavelength at 1,100 nm with <0.1% transmission in the reflection band and >98% transmission in the passband. The substrate is sapphire, and the films to be used have refractive indices of 2.5 and 1.7. The transition region should be smaller than 100 nm.
2. For the above filter, replace the constant-index materials with TiO$_2$ and Al$_2$O$_3$, and examine the change in performance. Then perform a numerical optimization to improve its performance.

3. For a certain application, we need a 99% reflection at 2,800 nm and 0% reflection at a wavelength of 1,650 nm on a CaF_2 substrate. Design the multilayer structure for this application using $\left(\frac{H}{2}L\frac{H}{2}\right)$ as the unit cell with Si and SiO_2 as the film materials. Plot the equivalent index of this structure as a function of normalized frequency G. Then find the layers that need to be added to the top and bottom of the stack such that the reflection at the substrate interface and the air interface can be reduced at 1,650 nm.

REFERENCES

1. Escoubas, L., Drouard, E. & Flory, F. Designing waveguide filters with optical thinfilm computational tools. *Optics Communications* **197**, 309–316, (October 2001).
2. Willey, R. R. Estimating the number of layers required and other properties of blocker and dichroic optical thin films. *Applied Optics* **35**, 4982–4986, (September 1996).
3. Tang, J. F. & Zheng, Q. Automatic design of optical thin-film systems—merit function and numerical optimization method. *The Journal of the Optical Society of America* **72**, 1522–1528, (November 1982).
4. Dobrowolski, J. A. & Kemp, R. A. Refinement of optical multilayer systems with different optimization procedures. *Applied Optics* **29**, 2876–2893, (July 1990).
5. Larouche, S. & Martinu, L. OpenFilters: Open-source software for the design, optimization, and synthesis of optical filters. *Applied Optics* **47**, C219. ISSN: 0003-6935 (May 2008).

9 Line-Pass Filters

9.1 SINGLE-CAVITY DESIGN

The methods for creating antireflectors and high reflectors were established in Chapters 3–7. In Chapter 8, we examined techniques for combining a high-reflection region with an adjacent antireflection region to create edge filters. This was done by examining the equivalent index η of the unit cell in the passband and inserting additional layers to create an antireflection in the passband while preserving the high reflection at the reference wavelength. In this chapter, we will examine methods for placing an antireflection region at the center of the high reflector's spectral band.

As we did with edge filters, we will begin with a high reflector based on the unit cell $\left(\frac{H}{2}L\frac{H}{2}\right)$. This structure will produce its maximum reflection at the reference wavelength. The goal is to turn the maximum reflection at the reference wavelength to a minimum reflection, without significantly affecting the remainder of the reflection spectra. We can do this if we can somehow turn the entire reflector into an absentee structure at the reference wavelength.

Consider, for example, placing a $\frac{H}{2}H$ layer on top of the reflector $\left(\frac{H}{2}L\frac{H}{2}\right)^N$:

$$\underbrace{\frac{H}{2}H}\ \ \underbrace{\frac{H}{2}LHLHLHLH\ldots\frac{H}{2}}_{\left(\frac{H}{2}L\frac{H}{2}\right)^N}. \tag{9.1}$$

The $\frac{H}{2}H$ will combine with the $\frac{H}{2}$ on the top of the reflector stack to become a HH absentee layer. If we place another L left of the $\frac{H}{2}H$,

$$L\frac{H}{2}H\ \ \underbrace{\frac{H}{2}LHLHLHLH\ldots\frac{H}{2}}_{\left(\frac{H}{2}L\frac{H}{2}\right)^N}, \tag{9.2}$$

that would combine with the L in the stack to become an LL absentee layer. Following this argument, we can show that

$$\left(\frac{H}{2}L\frac{H}{2}\right)^N H \left(\frac{H}{2}L\frac{H}{2}\right)^N \tag{9.3}$$

would result in the entire structure disappearing as absentee layers except for the two outer $\frac{H}{2}$ layers. In other words, at the reference wavelength, two reflector stacks $\left(\frac{H}{2}L\frac{H}{2}\right)^N$ separated by a quarter-wave H layer at the center collapses to just one H layer.

Therefore, the reflection at the reference wavelength will be determined only by the H layer and the underlying substrate. It is a relatively simple matter to add layers above and/or below the stacks to eliminate this reflection using one of many techniques that were discussed in Chapters 3 and 5. The most common method is to place two $\frac{H}{2}$ layers at the top and bottom of the stack to cancel out the resulting H layer and then apply antireflection coating to the substrate under it such that its effective reflectance index n_r is 1.0, such as:

$$\text{Substrate } (n_r = 1.0) \left| \frac{H}{2} \left(\frac{H}{2} L \frac{H}{2} \right)^N H \left(\frac{H}{2} L \frac{H}{2} \right)^N \frac{H}{2} \right. . \tag{9.4}$$

For example consider $n_1 = 2.5$, $n_2 = 1.5$, and $N = 5$ on an appropriately coated substrate such that its n_r is equal to 1.0 at the reference wavelength. The reflection spectrum of this structure computed using the transfer matrix method is shown in Figure 9.1. The zero-reflectance line at the center of the reflection band is clearly evident. The magnified plot in Figure 9.1b shows that the transmission line is about 2 nm wide.

While the above argument can be used to explain the high transmission line at the reference wavelength, it does not quite explain the high reflectance properties at all other wavelengths immediately adjacent to this transmission line. We can attempt to explain this by considering the effective reflectance index contours.

At $G = 1$, the structure given by (9.4) will describe a contour as shown in Figure 9.2a. The arc that belongs to the central H layer traces a very large circle, so only a small portion of it is shown on this plot. The contour starts at $n_r = 1.0$ and a right turning arc due to the first $H/2$ layer. Next, each unit cell of $\frac{H}{2} L \frac{H}{2}$ brings the index progressively closer toward the Herpin's equivalent index value of $j2.5$. The final value after three unit cells is $0.09 + j2.49$. Because its real value is so small, the next quarter-wave trace is very large, intercepting the real axis on the right hand side at very large values. Furthermore, this arc is nearly a complete circle even though it is only a quarter-wave film. This arises due to the nonuniform distribution of phase along the contour, as was discussed in Figure 5.7a in Chapter 5. The end point of this large arc is a mirror image about the real axis. As a result, the next three unit cells and the final $H/2$ film bring the end point exactly to the starting point. The

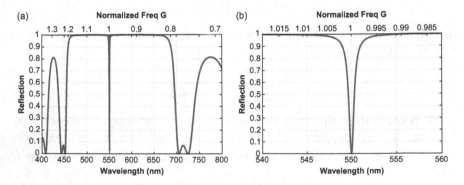

Figure 9.1 Reflection spectrum of $\frac{H}{2} \left(\frac{H}{2} L \frac{H}{2} \right)^5 H \left(\frac{H}{2} L \frac{H}{2} \right)^5 \frac{H}{2}$ with $n_1 = 2.5$, $n_2 = 1.5$, and substrate $n_r = 1.0$. (a) Full spectrum. (b) Closeup of the transmission line.

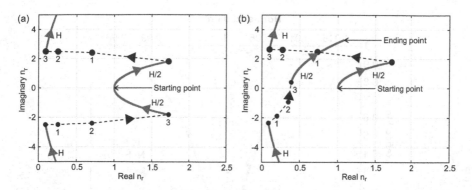

Figure 9.2 Effective reflection index contours of $\frac{H}{2}\left(\frac{H}{2}L\frac{H}{2}\right)^3 H\left(\frac{H}{2}L\frac{H}{2}\right)^3 \frac{H}{2}$ with $n_1 = 2.5$, $n_2 = 1.5$, and substrate $n_r = 1.0$ at $G = 1.0$ and $G = 1.01$. (a) At the reference wavelength ($G = 1$). (b) At $G = 1.01$.

starting and end point being the same satisfies the absentee condition, resulting in zero reflection.

If the frequency is slightly increased, as shown in Figure 9.2b, the ending point of the central H layer will become slightly displaced. The contour of the $\frac{H}{2}L\frac{H}{2}$, because of its large imaginary value, is extremely sensitive to the starting point. As a result, we see that the end point is much different than that in Figure 9.2a.

We could have also interpreted the structure in equation (9.3) as a resonant cavity enclosed by two reflectors. At the resonance wavelength, the cavity transmission will reach a maximum, and the reflection will reach a minimum, consistent with the absentee condition. A Fabry–Pérot cavity is a similar example, except in our case the cavity length is very small (quarter wave thick), and the mirror reflectivities and their phases are not constants but are a strong function of wavelength.

In these structures, the width of the transmission line is primarily controlled by the strength of the reflector, which in turn is determined by the number of repeating unit cells in the reflector stack ($N = 5$ in this example). A greater number of unit cells in the reflector stack will result in a sharper transmission peak. This peak transmission value will typically be a single point. While this may seem like a good thing in line filters, in practice any small deviations from this peak location can lead to significantly reduced transmission values. The deviations can arise from manufacturing tolerances of the filter or due to shifts in the laser wavelength that is being transmitted through the line filter. Additionally, it can also arise due to small changes in the angles of incidence on the filter. Therefore, it is often advantageous to create a flatter profile at the transmission peak to allow for some extra tolerance. One way to do this is by decreasing the number of layers that make up the high-reflection stack. For example, $\frac{H}{2}\left(\frac{H}{2}L\frac{H}{2}\right)^3 H\left(\frac{H}{2}L\frac{H}{2}\right)^3 \frac{H}{2}$ will have a broader transmission spectrum than $\frac{H}{2}\left(\frac{H}{2}L\frac{H}{2}\right)^5 H\left(\frac{H}{2}L\frac{H}{2}\right)^5 \frac{H}{2}$. The reflection spectra of both these structures are shown on Figure 9.3. However, a major disadvantage of reducing the number of layers is the loss of transition gradient on both sides of the high transmission peak. It is difficult to achieve a flatter peak with a very sharp gradient using a single-cavity

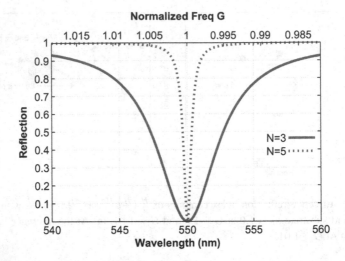

Figure 9.3 Comparison of the reflection spectra of $\frac{H}{2}\left(\frac{H}{2}L\frac{H}{2}\right)^N H \left(\frac{H}{2}L\frac{H}{2}\right)^N \frac{H}{2}$ line-pass filters for $N = 3$ and $N = 5$ using with $n_H = 2.5$ and $n_L = 1.5$ and substrate $n_r = 1.0$.

filter. One solution to overcome this problem is to use a coupled-cavity configuration, which will be discussed in Section 9.3 of this chapter.

While there was no difference between $\left(\frac{H}{2}L\frac{H}{2}\right)^N$ and $\left(\frac{L}{2}H\frac{L}{2}\right)^N$ with respect to their reflectivity, they do exhibit a difference in the width of the resonant transmission of the line filter. This arises due to the refractive index contrast between the cavity layer and the reflectors. At the reference wavelength, the equivalent index of the trilayer unit cell was jn_1 (equation (7.49) from Chapter 7). The refractive index contrast between the cavity and the reflector will be $n_H - jn_H$ in the case of $\left(\frac{H}{2}L\frac{H}{2}\right)^N$ and $n_L - jn_L$ in the case of $\left(\frac{L}{2}H\frac{L}{2}\right)^N$. The first is obviously larger in magnitude, which results in a stronger cavity resonance (or Q-factor). This is demonstrated in Figure 9.4.

The width of the reflection band around the transmission peak is primarily determined by the refractive index contrast between the high- and low-index films. This aspect was previously explored in detail in Chapter 7. A large index contrast will produce a broad reflection band.

At this point, it is useful to introduce an abbreviated notation to represent these cavity geometries. We will abbreviate $\left(\frac{H}{2}L\frac{H}{2}\right)^N$ as simply R_{NH}, where the subscripts stand for the number of unit cells in the reflector and the outside layer type (high index or low index). Therefore, the cavity geometry of Figure 9.1 can be written more compactly as $\frac{H}{2}R_{5H}HR_{5H}\frac{H}{2}$.

9.1.1 RESONANT-CAVITY ENHANCEMENT

An interesting aspect of these line filter structures is the field intensity. At the reference wavelength, the field intensity grows until it reaches a maximum value

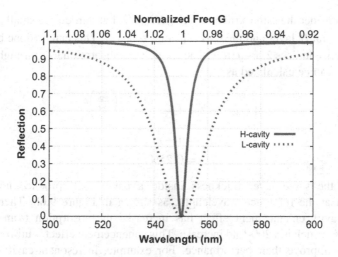

Figure 9.4 Comparison of the reflection spectra of the $\frac{H}{2}\left(\frac{H}{2}L\frac{H}{2}\right)^3 H\left(\frac{H}{2}L\frac{H}{2}\right)^3 \frac{H}{2}$ and $\frac{L}{2}\left(\frac{L}{2}H\frac{L}{2}\right)^3 L\left(\frac{L}{2}H\frac{L}{2}\right)^3 \frac{L}{2}$ line-pass filters using film indices of 2.5 and 1.5 and substrate $n_r = 1.0$.

at the central quarter-wave layer. This layer is also referred to as the defect layer (in the context of photonic crystals) or as the phase-shift layer. The effect is known as resonant-cavity enhancement, and any material effects specific to that central layer can be enhanced many fold [1,2]. This can be used to increase absorption in photodetectors [3] or to increase gain in optical amplifiers.

For example, the field intensity distribution in $\frac{H}{2}R_{8H}HR_{8H}\frac{H}{2}$, assuming $n_1 = 2.5$ and $n_2 = 1.5$, is shown in in Figure 9.5. The reference wavelength is 550 nm. We can see that the field intensity near the center is about 60 times greater than the incident field. However, the maximum field is not exactly in the central H layer; it is in the two adjacent $\frac{H}{2}$ layers. Therefore, it is convenient to consider the central HH half-wave layer as the region where the maximum field is present.

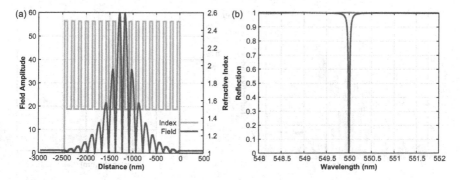

Figure 9.5 $\frac{H}{2}R_{8H}HR_{8H}\frac{H}{2}$ using $n_1 = 2.5$, $n_2 = 1.5$, and substrate $n_r = 1.0$. (a) Field distribution. (b) Reflection spectrum.

Next, consider the same structure $\frac{H}{2}R_{8H}HR_{8H}\frac{H}{2}$, but introduce a small absorption into the central H layer and the two adjacent $\frac{H}{2}$ layers. This can be done by making their refractive indices $2.5 - j10^{-3}$. The single-pass absorption loss through this half-wave layer can be calculated as

$$1 - e^{-2k_0 t n_i} = 1 - e^{-2\left(\frac{2\pi}{\lambda_0}\right)\left(\frac{\lambda_0}{2n_f}\right)n_i} \tag{9.5}$$

$$= 1 - e^{-2\pi\frac{n_i}{n_f}} \tag{9.6}$$

$$= 0.25\%. \tag{9.7}$$

However, the same layer thickness inside a cavity will produce nearly 50% absorption at the reference wavelength, as shown in Figure 9.6b. Therefore, the resonant-cavity enhancement effect has increased the absorption from 0.25% to nearly 50%, which is a 200-fold increase. This enhancement effect is utilized in many devices to improve their performance. For example, in resonant-cavity-enhanced (RCE) photodetectors, the absorption of a thin detector layer is improved by placing it inside a cavity. Similarly, in Vertical Cavity Surface Emitting Lasers (VCSELs), the optical gain from a thin layer is increased many fold by placing it inside a cavity.

Strictly speaking, we cannot replace the central quarter-wave film with a complex refractive index and expect the resonance condition and wavelength to remain unaltered. Even if the real part of the refractive index is the same, the imaginary part of the index will affect the antireflection (absentee) condition at the reference wavelength. As a result, the reflection at the reference wavelength will not be exactly zero, and the resonance wavelength will also shift (due to the central layer not having a precise $\frac{\pi}{2}$ phase). However, as long as the imaginary part is very small (as was the case in the previous example), its impact on the resonance condition will be minimal. If, however, the imaginary part of the refractive index is large, then we need to include its effect when designing the entire layer structure. We will defer this topic until Chapter 13, Section 13.2.3. The interested reader should read ahead to that

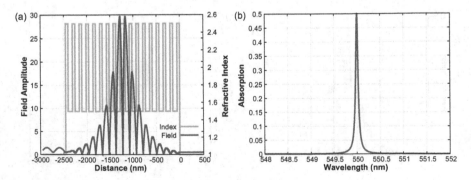

Figure 9.6 Field distribution and absorption spectrum of $\frac{H}{2}R_{8H}HR_{8H}\frac{H}{2}$ using $n_1 = 2.5$, $n_2 = 1.5$ and substrate $n_r = 1.0$. (a) Field distribution. (b) Absorption spectrum.

section. In the next section on VCSELs, we will continue with the assumption that the imaginary part of the central layer is sufficiently small that the resonance is not significantly affected.

9.2 VCSELs

VCSELs utilize a similar structure to enhance the optical gain in the central layers [4,5]. This enables lasing action to occur with optical gain present in only a very small region of the structure. Stated differently, the threshold for lasing is significantly reduced due to the resonant-cavity effect.

Typical VCSELs use GaAs and AlAs as the alternating layers. These films are different than the films typically used in general optical filters. They are semiconductor films grown epitaxially as a single crystal to preserve their electronic band structure because electrons and holes need to flow through these layers to induce population inversion and produce optical gain in the central region. The refractive index of GaAs is about 3.5 (H-layer), and AlAs (L-layer) is 2.9. The central HH region contains the gain medium, which is usually composed of multiple quantum wells (MQWs). Because the refractive index contrast between GaAs and AlAs is smaller compared to that in our previous examples, the number of unit cells required will be larger.

VCSELs typically contain about 15 or 20 unit cells for the mirrors. Figure 9.7 shows the reflection spectrum and the field distribution for $\frac{H}{2}R_{15H}HR_{15H}\frac{H}{2}$ using a reference wavelength of 1,300 nm. The substrate is GaAs, but an additional quarter-wave layer with an index $\sqrt{3.5} = 1.87$ has been inserted above the substrate to make the effective reflectance index of the substrate equal to 1.0. We can clearly see the sharp resonance at 1,300 nm. Additionally, the field distribution also illustrates the cavity resonance of this structure, with most of the energy concentrated in the HH layers at the center.

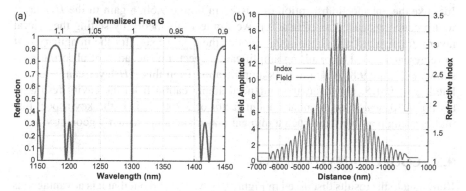

Figure 9.7 Reflection spectrum and field distribution of a symmetric VCSEL cavity consisting of 15 unit cells for each reflector, $\frac{H}{2}R_{15H}HR_{15H}\frac{H}{2}$, using GaAs and AlAs at a reference wavelength of 1,300 nm. The substrate is GaAs with an additional index matching layer. (a) Reflection Spectrum. (b) Field distribution at $\lambda = 1,300$ nm.

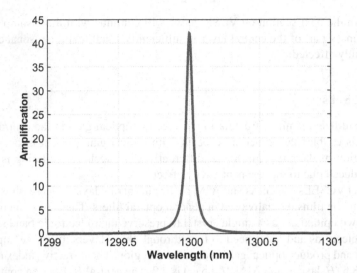

Figure 9.8 Optical amplification in $\frac{H}{2}R_{15H}HR_{15H}\frac{H}{2}$ using $n_1 = 3.5, n_2 = 2.9$, and a substrate with $n_1 = 3.5$ with an additional index matching layer. The central HH layer has an index of $3.5 + j10^{-2}$.

Furthermore, since the top mirror is used as the light output, the cavity is designed such that the top mirror has a slightly lower reflectivity than the bottom mirror. For example, $\frac{H}{2}$ R_{15H} H R_{12H} $\frac{H}{2}$ contains 15 unit cells on the bottom stack and 12 unit cells on the top. Additionally, lasers are not designed to filter light from the outside but rather filter the light originating from inside the cavity. The input is spontaneous emission in the gain region. As a result, optical modeling of lasers requires more considerations than what we have discussed so far. However, we can get an idea of the emission spectrum from a VCSEL by introducing gain in the central HH layer. Just like the enhanced absorption discussed in Figure 9.6b, a gain in the HH layer will produce enhanced amplification. We can test this theory by setting the central layer refractive index to $2.5 + j10^{-2}$ (notice the positive sign for the imaginary part to represent gain). The resulting amplification spectrum (negative of absorption) is shown in Figure 9.8. A single-pass amplification from this HH layer can be calculated to be 0.018. However, the enhanced amplification from the cavity results in a gain of 40, which is an enhancement factor of 2,200. This is the key aspect that makes lasing possible from such a thin gain layer and a microcavity geometry.

9.3   COUPLED-CAVITY DESIGN

Returning to the results discussed in Figure 9.3, we pointed out that it is advantageous to create a flatter profile at the transmission peak without compromising the transition gradient. One method that is used to widen the transmission line without compromising the transition gradient is to create two closely spaced transmission peaks next to each other. This can be achieved by coupling two cavities together, which produces a well-known splitting effect of the resonance. The coupling between two identical cavities will create two resonant frequencies slightly offset from each other, with the magnitude of the offset determined by the strength of the coupling.

For example, assuming a single resonator has a normalized resonance frequency of G_0, a system consisting of two of these resonators coupled together will exhibit two resonance frequencies of $G_0 + \Delta G$ and $G_0 - \Delta G$, where ΔG will be determined by the coupling strength. If the resonators are closely spaced such that the fields in the individual resonators interact strongly, ΔG will be large. If they are far apart with little interaction, ΔG will be small. In the extreme case where the resonators are infinitely apart such that there is no interaction between them, then ΔG will be zero, and the resonance will be at G_0. This effect is qualitatively illustrated in Figure 9.9. Alternatively, we can also view this like a system of two atoms (or electrons in a quantum well) separated by an energy barrier. The splitting of their frequencies (or energies) is a well-known phenomenon in many areas of science.

Expanding on the structure described by equation (9.3), we can create a two-cavity coupled system by

$$\left(\frac{H}{2}\right) \underbrace{\left(\frac{H}{2}L\frac{H}{2}\right)^N}_{\text{Reflector}-R_{NH}} H \underbrace{\left(\frac{H}{2}L\frac{H}{2}\right)^N}_{\text{Reflector}-R_{NH}} H \underbrace{\left(\frac{H}{2}L\frac{H}{2}\right)^N}_{\text{Reflector}-R_{NH}} \left(\frac{H}{2}\right). \qquad (9.8)$$

In this system, the two cavities are separated by three reflectors. The coupling strength between the two cavities will be determined by the reflection properties of the central reflector. We should be able to verify by expanding the expression in equation (9.8) that the structure does not collapse to an absentee structure at the reference wavelength. As a result, the reflection at the reference wavelength will not be zero. The zero reflection points actually move outward on both sides of the reference wavelength due to the splitting effect.

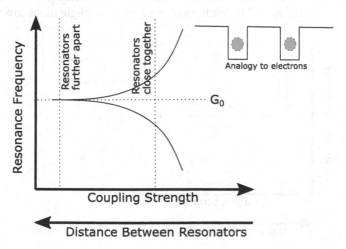

Figure 9.9 Illustration of the effect of coupling between two resonators.

A three-cavity coupled system (using four reflectors) such as

$$\left(\frac{H}{2}\right)\underbrace{\left(\frac{H}{2}L\frac{H}{2}\right)^N}_{R_{NH}}H\underbrace{\left(\frac{H}{2}L\frac{H}{2}\right)^N}_{R_{NH}}H\underbrace{\left(\frac{H}{2}L\frac{H}{2}\right)^N}_{R_{NH}}H\underbrace{\left(\frac{H}{2}L\frac{H}{2}\right)^N}_{R_{NH}}\left(\frac{H}{2}\right)\qquad(9.9)$$

will restore the symmetry such that it collapses to an absentee structure at the reference wavelength. In other words, an odd number of coupled cavities will have zero reflection at the reference wavelength. The splitting effect due to coupling between three cavities is illustrated in Figure 9.10.

Figure 9.11 shows the computed reflection spectra of the two-cavity and three-cavity coupled systems using five units cells in each reflector. The two-cavity structure shows two adjacent transmission peaks. The three-cavity structure shows three transmission peaks, and the central peak is coincident with the reference wavelength, consistent with the illustration in Figure 9.10. Nevertheless, the end result from these structures is a multiline filter – not a broadened single-line filer. This arises due to the high-reflection regions between the resonance points. We can improve this design by bringing the resonances closer together. Reducing the coupling strength (by increasing the reflectivity between the cavities) will reduce the spacing frequency between the resonances, and increasing the coupling strength (by reducing the reflectivity between cavities) will push the resonant frequencies further apart. Considering the two-cavity coupled system, the spacing can be reduced by increasing the reflectivity of the central reflector. For example, we could employ a configuration such as

$$\left(\frac{H}{2}\right)\underbrace{\left(\frac{H}{2}L\frac{H}{2}\right)^3}_{R_{3H}}H\underbrace{\left(\frac{H}{2}L\frac{H}{2}\right)^5}_{R_{5H}}H\underbrace{\left(\frac{H}{2}L\frac{H}{2}\right)^3}_{R_{3H}}\left(\frac{H}{2}\right).\qquad(9.10)$$

The spacing between peaks can be further reduced by increasing the reflectivity of the central reflector to 7. The performance of these configurations are shown in

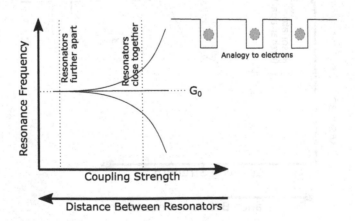

Figure 9.10   Illustration of the effect of coupling between three resonators.

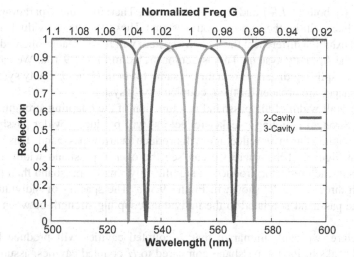

Figure 9.11 Reflection spectrum of the coupled system $\frac{H}{2}R_{3H}HR_{3H}HR_{3H}\frac{H}{2}$ and the three-cavity coupled system $\frac{H}{2}R_{3H}HR_{3H}HR_{3H}HR_{3H}\frac{H}{2}$ using $n_1 = 2.5$ and $n_2 = 1.5$ and substrate $n_r = 1.0$.

Figure 9.12a. We can see that $\frac{H}{2}R_{3H}HR_{7H}HR_{3H}\frac{H}{2}$ may be a good starting point to optimize further. Figure 9.12b shows the numerically optimized result using $\frac{H}{2}R_{3H}HR_{7H}HR_{3H}\frac{H}{2}$ as the starting design.

The choice of $\left(\frac{H}{2}L\frac{H}{2}\right)$ versus $\left(\frac{L}{2}H\frac{L}{2}\right)$ also plays a role in the overall behavior of the coupled system. As discussed earlier in this chapter, the H cavity system employing $\left(\frac{H}{2}L\frac{H}{2}\right)$ as reflectors will exhibit a stronger refractive index contrast compared to an L cavity system employing $\left(\frac{L}{2}H\frac{L}{2}\right)$. This results in the H cavity exhibiting a stronger and sharper resonance. However, the coupling between cavities is dictated only by the total reflectivity between the cavities, which is a function of $N\theta$. This is

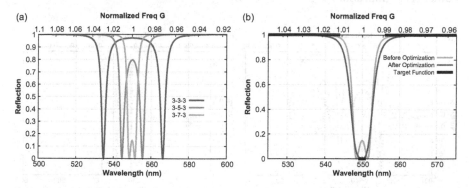

Figure 9.12 Reflection spectrum of the different coupled systems with $n_1 = 2.5$ and $n_2 = 1.5$ and substrate $n_r = 1.0$. (a) $\frac{H}{2}R_{3H}HR_{3H}HR_{3H}\frac{H}{2}$ vs $\frac{H}{2}R_{3H}HR_{5H}HR_{3H}\frac{H}{2}$ vs $\frac{H}{2}R_{3H}HR_{7H}HR_{3H}\frac{H}{2}$. (b) $\frac{H}{2}R_{3H}HR_{7H}HR_{3H}\frac{H}{2}$ after numerical optimization.

the same for both $\left(\frac{H}{2}L\frac{H}{2}\right)$ and $\left(\frac{L}{2}H\frac{L}{2}\right)$ reflectors. Therefore, the separation between the split lines will be the same in both systems. However, the individual resonance lines will exhibit a difference: the L cavity system will have broader line widths compared to the H cavity system. This is demonstrated in Figure 9.13a. We can clearly see that the splitting distances remain the same for both H and L cavity systems, but the line shapes are broader for the L coupled cavity system.

The overall width of the passband is a function of the coupling strength between cavities. Stronger coupling (weaker reflectors) will produce a wider passband, and weaker coupling (stronger reflectors) will produce a narrow passband. This is demonstrated in Figure 9.13b, where we can see the overall passband width using five unit cells in each reflector produces a significantly smaller passband than the structure with three unit cells shown in Figure 9.13a. The spacing of individual peaks inside the passband is related to the individual coupling strength between adjacent cavities.

Therefore, we can summarize that L coupled cavities will produce less pronounced peaks inside the passband compared to H coupled cavities, assuming N is the same for both, although this comes at the expense of the transition edge gradient, while increasing N has the effect of reducing the overall passband width.

As a result, we have a number of independent parameters we can utilize to tweak the performance of the design. This includes coupling between cavities, choice of H versus L cavities, and the number of unit cells in the reflectors, as well as the distribution of N across the cavities. Unfortunately, there is no simple algorithm for synthesizing these designs. Some trial and error and some intuition is necessary to achieve the desired response.

Figure 9.14 shows the response of a five-cavity coupled system

$$\frac{L}{2}R_{2L}\ L\ R_{4L}\ L\ R_{5L}\ L\ R_{5L}\ L\ R_{4L}\ L\ R_{2L}\frac{L}{2}. \tag{9.11}$$

This structure has a nearly perfect broadened line-pass characteristic, even without numerical optimization.

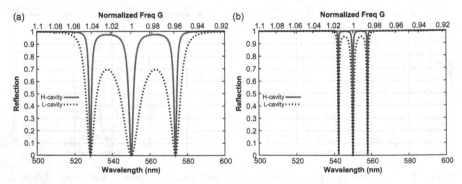

Figure 9.13 Three-cavity coupled system using H and L cavities with refractive indices of 2.5 and 1.5 and a substrate $n_r = 1.0$. (a) $N = 3$ and (b) $N = 5$.

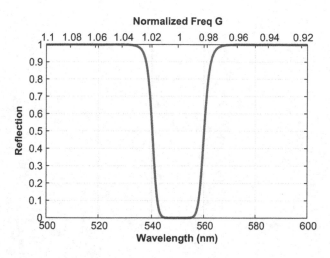

Figure 9.14 Reflection spectra of five-cavity coupled system $\frac{L}{2}R_{2L}LR_{4L}LR_{5L}LR_{5L}LR_{4L}LR_{2L}\frac{L}{2}$ *without* numerical optimization using $n_1 = 1.5$ and $n_2 = 2.5$ and substrate $n_r = 1.0$.

9.4 PROBLEMS

1. Design a three-cavity line-pass filter for the YAG laser wavelength of 1,064 nm, using TiO_2 and SiO_2 as the high-index material and low-index material, respectively, on a SiO_2 substrate.

2. Consider a single-cavity 632 nm line-pass filter using MgF_2 and ZnS as the layers. The central ZnS H layer is doped such that its imaginary part of the index is 1.0×10^{-4}. Determine the number of unit cells required in this symmetric cavity such that the RCE absorption in the central H layer is 10%.

3. Using the structure described in equation (9.11) (Figure 9.14), modify this design to create a line-pass filter at a wavelength of 4.0 μm using Si and SiO as the materials. The substrate is ZnS, and it has to be appropriately coated to allow full transmission at the reference wavelength.

REFERENCES

1. Ünlü, M. S. & Strite, S. Resonant cavity enhanced photonic devices. *Journal of Applied Physics* **78**, 607–639 (July 1995).
2. Schubert, E. F., Wang, Y.-H., Cho, A. Y., Tu, L.-W. & Zydzik, G. J. Resonant cavity light-emitting diode. *Applied Physics Letters* **60**, 921–923 (Feburary 1992).
3. Kishino, K. et al. Resonant cavity-enhanced (RCE) photodetectors. *IEEE Journal of Quantum Electronics* **27**, 2025–2034 (1991).
4. Michalzik, R. *VCSELs: Fundamentals, Technology and Applications of Vertical-Cavity Surface-Emitting Lasers. Springer Series in Optical Sciences*, Vol. 166, (Springer, New York, 2013). ISBN: 9783642249853.
5. Iga, K. Vertical-Cavity Surface-Emitting Laser (VCSEL). *Proceedings of the IEEE* **101**, 2229–2233 (October 2013).

Figure 3.13 ...

4.3 PROBLEMS

REFERENCES

10 Bandpass Filters

10.1 BANDPASS FILTERS BY COMBINING TWO EDGE FILTERS

The simplest type of bandpass filter is one that consists of two edge filters. It is possible to construct a long-pass filter and combine this with a short-pass filter such that the high-reflection band of the short-pass filter falls inside the high-transmission side of the long-pass filter. For example, we could use an optimized long-pass filter based on $\left(\frac{H}{2}L\frac{H}{2}\right)^N$ and place this directly above a short-pass filter based on $\left(\frac{L}{2}L\frac{L}{2}\right)^N$ as shown in Figure 10.3a. The reference wavelength of the long-pass filter has to be shorter than the reference wavelength of the short-pass filter. Figure 10.1 shows an example of this filter configuration. Figure 10.1a shows a long-pass filter using $\left(\frac{H}{2}L\frac{H}{2}\right)^{10}$ at a reference wavelength of 450 nm using $n_1 = 2.5$ and $n_2 = 1.5$. The substrate was assumed to have an effective reflectance index n_r equal to 1.0. This structure was numerically optimized to achieve a transition at $\lambda = 550$ nm. Figure 10.1b shows a short-pass filter using $\left(\frac{L}{2}H\frac{L}{2}\right)^{10}$ at a reference wavelength of 900 nm and numerically optimized to achieve a transition at $\lambda = 750$ nm.

Figure 10.2 shows the effect of simply stacking both of these filters such as

$$\left[\left(\frac{H}{2}L\frac{H}{2}\right)^{10}_{450 \text{ nm}}\right]\left[\left(\frac{L}{2}H\frac{L}{2}\right)^{10}_{900 \text{ nm}}\right]. \qquad (10.1)$$

The net effect produces a response that looks like a simple product of the two individual transmittances. However, caution should be exercised not to allow the

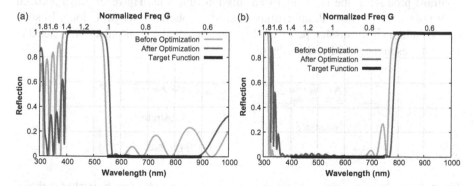

Figure 10.1 Separate long- and short-pass filters using film indices of 2.5 and 1.5. (a) Optimized $\left(\frac{H}{2}L\frac{H}{2}\right)^{10}$ long-pass filter using a reference wavelength of 450 nm. (b) Optimized $\left(\frac{L}{2}H\frac{L}{2}\right)^{10}$ short-pass filter using a reference wavelength of 900 nm.

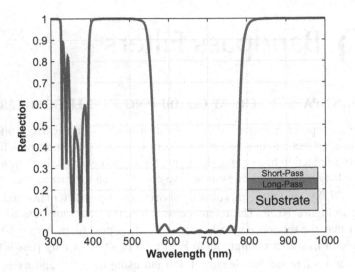

Figure 10.2   Stacked long-pass and short-pass filters.

high-reflection zones of both filters to overlap, because that could result in high-transmission peaks appearing inside the reflection zones due to cavity resonances between the two high reflectors. This means that the starting long-pass and short-pass filters should have fairly wide-pass bands.

If an overlap between the two high reflectors is unavoidable, an alternative approach is to physically decouple the two filters from each other to prevent any resonances from occurring between the two reflectors. This can be done by placing the two filters on opposite sides of the substrate as shown in Figure 10.3b. Since the substrate is generally much thicker than the coherence length of broadband light sources, this effectively decouples the two filters making the combined performance a simple product of the individual transmission functions. However, this approach may not work with laser illuminations where the coherence length is long. In such cases, multiple transmission peaks could appear inside the high-reflection zones due to cavity resonances within the substrate.

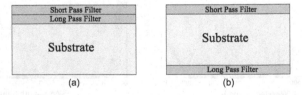

Figure 10.3   As long as high-reflection regions do not overlap, filters can be stacked as shown on the left. Alternatively, they can be placed on opposite sides of the substrate to reduce coupling between the filters. (a) Both filters on the same side of the substrate. (b) Filters on opposite sides of the substrate.

10.2   COUPLED-CAVITY BANDPASS FILTERS

Another approach for building bandpass filters is to utilize coupled-cavity filters as the basic elements. Line-pass filters based on coupled cavities were examined in Chapter 9, where we saw that coupling cavities together will result in the splitting of the high-transmission (or low-reflection) peaks. Two coupled cavities will result in two transmission peaks, three coupled cavities will result in three transmission peaks, and so on. The examples we studied in Chapter 9 all had very narrow transmission bands (consistent with the definition of line-pass filters). One way to open up the transmission band is to use very strongly coupled cavities (i.e., using very weak reflectors between cavities). To prevent peaks from appearing between these resonances, we can select L as the cavity material.

The cavity structure we introduced in Chapter 9 was

$$\left(\frac{L}{2}\right) \underbrace{\left(\frac{L}{2}H\frac{L}{2}\right)^N}_{\text{Reflector}-R_{NL}} L \underbrace{\left(\frac{L}{2}H\frac{L}{2}\right)^N}_{\text{Reflector}-R_{NL}} \left(\frac{L}{2}\right). \tag{10.2}$$

The weakest possible reflector we can create is by selecting $N = 1$, which becomes $\frac{L}{2}R_{1L}LR_{1L}\frac{L}{2}$. This can also be expanded as

$$LH\,LL\,HL. \tag{10.3}$$

Because it is a weak cavity, it will behave as a poor line-pass filter; i.e., it will have gently sloping gradients very similar to the single-layer antireflection designs. The calculated reflection spectrum of this structure is shown in Figure 10.4a. We can see that the transmission peak (minimum reflection) occurs at the reference wavelength of 550 nm with higher reflection bands on either sides, consistent with a weak cavity resonance.

Nevertheless, we can improve the performance of this bandpass filter by extending this structure to a multicavity coupled system. Consider a structure consisting of two coupled cavities, such as $\frac{L}{2}R_{1L}LR_{1L}LR_{1L}\frac{L}{2}$:

$$\left(\frac{L}{2}\right) \underbrace{\left(\frac{L}{2}H\frac{L}{2}\right)}_{R_{1L}} L \underbrace{\left(\frac{L}{2}H\frac{L}{2}\right)}_{R_{1L}} L \underbrace{\left(\frac{L}{2}H\frac{L}{2}\right)}_{R_{1L}} \left(\frac{L}{2}\right). \tag{10.4}$$

In expanded notation, this is really

$$LH\,LL\,H\,LL\,HL. \tag{10.5}$$

The reflection spectrum of this structure is shown in Figure 10.4b. We can see that the slopes of the transitions have improved, and the reflection outside the passband has also increased from 60% to 80%. This concept can be carried further to include a large number of coupled cavities to produce very-well-behaved bandpass filters. This is shown in Figure 10.5 for a number of coupled cavities ranging from 1 to 10.

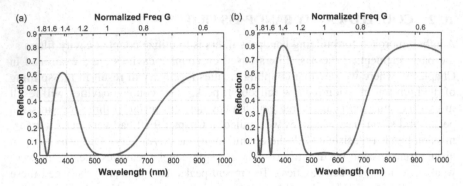

Figure 10.4 Single- and double-cavity bandpass filters with $n_1 = 1.5$ and $n_2 = 2.5$ and substrate $n_r = 1.0$. (a) Single cavity $- \frac{L}{2}R_{1L}LR_{1L}\frac{L}{2}$. (b) Double cavity $- \frac{L}{2}R_{1L}LR_{1L}LR_{1L}\frac{L}{2}$.

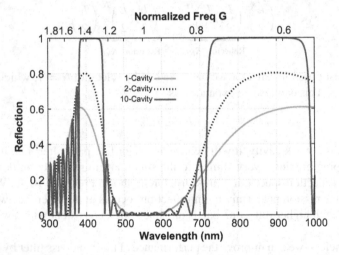

Figure 10.5 Reflection spectrum of multiple coupled-cavity system, ranging from one to ten cavities using R_{1L} with $n_1 = 1.5$ and $n_2 = 2.5$ and substrate $n_r = 1.0$.

We can note that the edges of the passband become much steeper as the number of cavities is increased, while the width of the passband remains the same.

Considering ten coupled cavities, we could use that as our baseline to numerically optimize the performance and to suppress the ripples inside the passband. The optimized result is shown in Figure 10.6.

Additionally, instead of using $\left(\frac{L}{2}H\frac{L}{2}\right)$ as the unit cell for the cavity reflectors, we could have chosen $\left(\frac{H}{2}L\frac{H}{2}\right)$ as the unit cell for the reflectors. This would result in H becoming the cavity instead of L, and equation (10.2) would become

$$\left(\frac{H}{2}\right) \underbrace{\left(\frac{H}{2}L\frac{H}{2}\right)}_{\text{Reflector}-R_{1H}} H \underbrace{\left(\frac{H}{2}L\frac{H}{2}\right)}_{\text{Reflector}-R_{1H}} \left(\frac{H}{2}\right). \tag{10.6}$$

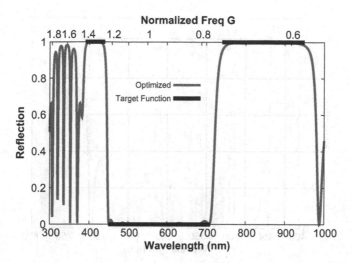

Figure 10.6 Numerically optimized ten-cavity filter using $n_1 = 1.5$ and $n_2 = 2.5$ and substrate $n_r = 1.0$ (using the ten-cavity result in Figure 10.5 as the baseline).

As discussed previously, the refractive index contrast between the cavity and the reflector will be larger with R_{NH} compared with R_{NL}. As a result, the field will be more strongly confined within the cavities with R_{NH}, resulting in weaker coupling between the cavities. This would result in the individual peaks inside the transmission band being distinctly separated more with R_{NH} than with R_{NL}.

A ten-cavity filter using R_{1H} is shown in Figure 10.7. We can clearly see that the passband has very strong peaks. In fact, there are ten peaks, corresponding to the

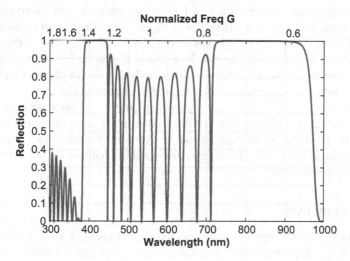

Figure 10.7 Ten-cavity filter using R_{1H} as the unit cell of the reflector, $n_1 = 2.5$, $n_2 = 1.5$, and substrate $n_r = 1.0$.

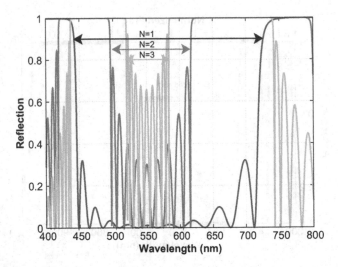

Figure 10.8 Three different structures of ten coupled-cavity bandpass filters using a different number of unit cells in each reflector to demonstrate the effects of the number of unit cells on the passband width.

number of cavities in the structure. The high-reflection regions between the transmission peaks arise due to the smaller coupling between the cavities.

To a large extent, the width of the whole passband is a function of the individual cavity strengths. For the previous examples, we chose $N = 1$ to produce a weak cavity and to give it a broad passband. A narrower passband can be designed by increasing the strength of each cavity. This is done by choosing a larger number of unit cells in each reflector, such as $N = 2$ or $N = 3$.

For example, the performance of a ten-cavity bandpass filter using R_{NL} as the reflectors with $N = 1, 2$, and 3 is shown in Figure 10.8. The refractive index values are $n_1 = 1.5$, $n_2 = 2.5$, and $n_r = 1.0$ as before. For $N = 1$, the width of the passband is about 250 nm. For $N = 2$, the width becomes 40 nm, and for $N = 3$, the width narrows down to 20 nm. Unfortunately, this method does not allow one to achieve arbitrary passband widths because N is limited to integer values.

Furthermore, in cases where $n_s \neq n_a$, we can employ the same strategy that was discussed in Chapter 9 by inserting an extra layer to match n_s to n_a at the reference wavelength. However, some care should be exercised because this matching only works for a limited wavelength range, as discussed in Chapters 3 and 5 in the context of antireflection designs.

10.3 PROBLEMS

1. Using Ge and an appropriately selected second material, design a coupled-cavity bandpass filter that transmits the mid-wave infrared (MWIR) spectral window of 3.0–4.0 μm. Then, numerically optimize this design to reduce the ripples inside the passband.

FURTHER READING

Goury, C. et al. Design and realization of multispectral bandpass filters for space applications in *International Conference on Space Optics—ICSO 2018* (eds Sodnik, Z., Karafolas, N. & Cugny, B.) **11180** (SPIE, 2019), 2969–2978.

Macleod, H. A. *Thin-Film Optical Filters (Series in Optics and Optoelectronics)*, (CRC Press, Boca Raton, FL, 2017). ISBN: 1138198242.

11 Thin-Film Designs for Oblique Incidence

11.1 ANGLE OF INCIDENCE ON THE SPECTRAL PERFORMANCE OF A FILTER

Even though the majority of optical coatings are designed for normal incidence, it is important to evaluate how they will perform for off-normal incidence. In addition, there are also applications where the incident beam is required to be at a specific angle, such as, for example, a dielectric mirror that has to operate at only 45° from normal. Therefore, it is important to develop a systematic method to design thin-film structures for oblique incidence. Additionally, the polarization of the incident beam has to be considered under oblique incidence. Under normal incidence, both transverse electric (TE) and transverse magnetic (TM) polarizations exhibit identical properties. However, under oblique incidence, each polarization will behave differently, unless specific measures are taken to create polarization-insensitive designs [1,2].

Figure 11.1 shows an example of a line-pass filter under different angles of incidence. The filter was designed for normal incidence using a reference wavelength of 550 nm. We can see that the effect of increasing angles is to produce a blue shift of the resonance wavelength. However, this is not a simple lateral shift; if it were, it would

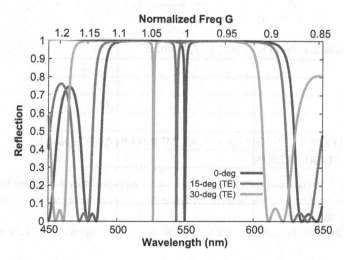

Figure 11.1 Performance of a line-pass filter using $\frac{H}{2}R_{8H}HR_{8H}\frac{H}{2}$ with $n_1 = 2.0$, $n_2 = 1.5$ and substrate $n_r = 1.0$, designed for normal incidence at a reference wavelength of 550 nm.

145

be a simple matter to account for it in the design. The blue shift is accompanied by spectral distortion and polarization sensitivity that can degrade the performance as the angle is increased. Therefore, in order to design a filter for a specific angle of incidence, we need to develop a systematic understanding of the origins of these effects. This is the subject of this chapter.

11.2   CONTINUITY EQUATIONS AND ANGLE OF INCIDENCE (TE)

The principles of oblique incidence are based on the field continuity equations discussed in Chapter 1. For TE incidence, we saw that the continuity of the fields and their derivatives across a dielectric interface at z_1 with refractive indices of n_a and n_f can be written as

$$e^{-jk_0\sqrt{n_a^2-n_a^2\sin^2(\theta_a)}z_1} + re^{+jk_0\sqrt{n_a^2-n_a^2\sin^2(\theta_a)}z_1} = te^{-jk_0\sqrt{n_f^2-n_a^2\sin^2(\theta_a)}z_1} \qquad (11.1)$$

and

$$\sqrt{n_a^2-n_a^2\sin^2(\theta_a)}e^{-jk_0\sqrt{n_f^2-n_a^2\sin^2(\theta_a)}z_1} - \sqrt{n_a^2-n_a^2\sin^2(\theta_a)}re^{+jk_0\sqrt{n_a^2-n_a^2\sin^2(\theta_a)}z_1}$$
$$= \sqrt{n_f^2-n_a^2\sin^2(\theta_a)}te^{-jk_0\sqrt{n_f^2-n_a^2\sin^2(\theta_a)}z_1}.$$
$$(11.2)$$

We can rewrite these equations by lumping the exponential factors and defining two equivalent indices

$$n_a^z = \sqrt{n_a^2-n_a^2\sin^2\theta} = n_a\cos\theta \qquad (11.3)$$

$$n_f^z = \sqrt{n_f^2-n_a^2\sin^2\theta}. \qquad (11.4)$$

As a result, equations (11.1) and (11.2) become

$$e^{-jk_0n_a^zz_1} + re^{+jk_0n_a^zz_1} = te^{-jk_0n_f^zz_1} \qquad (11.5)$$
$$n_a^ze^{-jk_0n_a^zz_1} - n_a^zre^{+jk_0n_a^zz_1} = n_f^zte^{-jk_0n_f^zz_1}. \qquad (11.6)$$

11.3   REFLECTION FROM A SINGLE INTERFACE FOR TE POLARIZATION

In Chapter 3, we derived the reflection across a single interface for normal incidence. We can follow the same methods to rederive the expression for the more general case of oblique incidence.

Assuming the dielectric interface z is at the origin, equations (11.5) and (11.6) become

$$1 - r = t \qquad (11.7)$$

$$n_a^z + n_a^zr = n_f^zt. \qquad (11.8)$$

Solving for r from the above equations results in

$$r = \frac{n_f^z - n_a^z}{n_f^z + n_a^z}. \tag{11.9}$$

In other words, the reflection from a dielectric interface for TE polarization is exactly the same as that in the normal-incidence case except that the refractive indices n_a, n_f etc. have been replaced by n_a^z, n_f^z, etc. Therefore, equivalent indices defined in equations (11.3) and (11.4) can be used as the equivalent indices of all materials in the structure under oblique incidence for the TE-polarized wave.

11.4 BEHAVIOR OF n^z WITH INCIDENT ANGLE

As shown previously, for all practical purposes, we can replace the refractive index of the material with n^z for TE polarization and treat the problem like a normal-incidence case. Therefore, it is instructive to examine how the values of n^z behave as a function of incidence angle. The calculated values of n_f^z are shown in Figure 11.2 for the incident medium (refractive index of 1.0) and two film materials with refractive indices of 1.5 and 2.5. We can see that all of the n_f^z values decline with increasing incident angles. This effect can be viewed similar to material dispersion with wavelength. Furthermore, materials with a lower refractive index experience a larger change than those with a higher refractive index. This is a useful fact because designs employing materials with high refractive indices will be relatively less sensitive to incident angles. Additionally, we can also observe that the curves diverge away from each other as the incident angle increases. This implies that the TE reflection will generally increase in magnitude with increasing angle, reaching 100% reflection at 90° incidence.

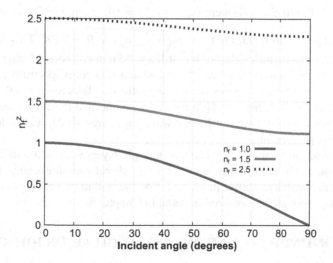

Figure 11.2 Calculated values of n_f^z for $n_f = 1.0$, 1.5, and 2.5.

11.5 SINGLE-LAYER ANTIREFLECTION FOR TE INCIDENCE

We can use the concepts developed in the previous sections to design antireflection films to operate at a specific incidence angle of θ. The process for designing this for a substrate with a refractive index of n_s is as follows.

For TE polarization, we derived expressions for n^z which can be used as direct replacement for the material index values. These were

$$n_s^z = \sqrt{n_s^2 - n_a^2 \sin^2 \theta} \tag{11.10}$$

$$n_f^z = \sqrt{n_f^2 - n_a^2 \sin^2 \theta} \tag{11.11}$$

$$n_a^z = \sqrt{n_a^2 - n_a^2 \sin^2 \theta} = n_a \cos \theta. \tag{11.12}$$

The required condition for antireflection using a single film was derived in Chapter 3, and it was shown that $n_f = \sqrt{n_s n_a}$. In this case, the required refractive index will be determined by n_s^z and n_a^z. Therefore, $n_f^z = \sqrt{n_s^z n_a^z}$. The thickness of the film will be $t_f = \frac{\lambda_0}{4n_f^z}$.

If we assume $n_s = 1.5$, $\theta = 45°$, and $\lambda_0 = 550$ nm, we can get the following values for the replacement indices:

$$n_s^z = 1.32 \tag{11.13}$$

$$n_a^z = 0.71 \tag{11.14}$$

$$n_f^z = \sqrt{n_s^z n_a^z} = 0.97. \tag{11.15}$$

The required film thickness is therefore $t_f = \frac{\lambda_0}{4n_f^z} = 142.2$ nm. Finally, we can convert n_f^z to n_f and get the film index to be $n_f = \sqrt{n_f^{z2} + n_a^2 \sin^2 \theta} = 1.198$. This is obviously too small for realistic materials, but it is still a valid numeric result. Using a fictitious material with this refractive index, the calculated reflection spectrum is shown in Figure 11.3. As expected, we can see that the reflection falls to zero at the reference wavelength at 45° incidence for TE polarization. At normal incidence, the minimum reflection point moves toward a longer wavelength (red-shift), with a less-perfect antireflection performance.

As discussed in Chapter 5, we can use multiple layers with materials with higher refractive indices to achieve this same condition, thereby circumventing the need to find a material with a refractive index of 1.198. We will not repeat those details here because they were already covered in detail in Chapter 5.

11.6 CONTINUITY EQUATIONS AND ANGLE OF INCIDENCE (TM)

For TM incidence, the continuity of the fields and their derivatives were derived in Chapter 1 as

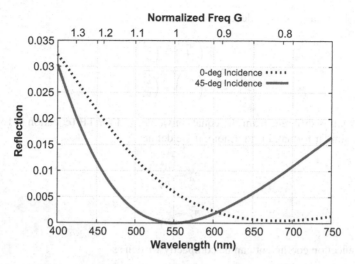

Figure 11.3 Reflection spectrum for TE polarization from a single-layer antireflection layer designed for 45° incidence for a reference wavelength of 550 nm.

$$\frac{1}{n_a}\sqrt{n_a^2 - n_a^2 \sin^2(\theta_a)}e^{-jk_0\sqrt{n_a^2-n_a^2\sin^2(\theta_a)}z_1} - \frac{1}{n_a}\sqrt{n_a^2 - n_a^2 \sin^2(\theta_a)}re^{+jk_0\sqrt{n_a^2-n_a^2\sin^2(\theta_a)}z_1}$$

$$= \frac{1}{n_f}\sqrt{n_f^2 - n_a^2 \sin^2(\theta_a)}te^{-jk_0\sqrt{n_f^2-n_a^2\sin^2(\theta_a)}z_1} \qquad (11.16)$$

and

$$n_a e^{-jk_0\sqrt{n_a^2-n_a^2\sin^2(\theta_a)}z_1} + n_a re^{+jk_0\sqrt{n_a^2-n_a^2\sin^2(\theta_a)}z_1}$$

$$= n_f t e^{-jk_0\sqrt{n_f^2-n_a^2\sin^2(\theta_a)}z_1}. \qquad (11.17)$$

Using the equivalent indices defined in equations (11.3) and (11.4), these equations can be simplified as

$$\frac{n_a^z}{n_a}e^{-jk_0 n_a^z z_1} - \frac{n_a^z}{n_a}re^{+jk_0 n_a^z z_1} = \frac{n_f^z}{n_f}te^{-jk_0 n_f^z z_1} \qquad (11.18)$$

$$n_a e^{-jk_0 n_a^z z_1} + n_a re^{+jk_0 n_a^z z_1} = n_f te^{-jk_0 n_f^z z_1}. \qquad (11.19)$$

11.7 REFLECTION FROM A SINGLE INTERFACE FOR TM POLARIZATION

Following the same approach as TE, we can use equations (11.18) and (11.19) to get the expressions for TM polarization. These are

$$\frac{n_a^z}{n_a}(1-r) = \frac{n_f^z}{n_f}t \qquad (11.20)$$

$$n_a + n_a r = n_f t, \qquad (11.21)$$

from which we can get the reflection coefficient as

$$r = \frac{\left(\frac{n_f^2}{n_f^z}\right) - \left(\frac{n_a^2}{n_a^z}\right)}{\left(\frac{n_f^2}{n_f^z}\right) + \left(\frac{n_a^2}{n_a^z}\right)}. \tag{11.22}$$

This equation is different from the equation we derived for TE (equation (11.9)), but it still has a similar format. Furthermore, if we define

$$n_a^{z\prime} = \frac{n_a^2}{n_a^z} \tag{11.23}$$

$$n_f^{z\prime} = \frac{n_f^2}{n_f^z}, \tag{11.24}$$

the TM reflection coefficient can be compactly written as

$$r = \frac{n_f^{z\prime} - n_a^{z\prime}}{n_f^{z\prime} + n_a^{z\prime}}. \tag{11.25}$$

This expression is identical in format to the TE reflection coefficient in equation (11.9) except the refractive indices n_a^z, n_f^z, etc. have been replaced with $n_a^{z\prime}$, $n_f^{z\prime}$, etc. In other words, $n_f z\prime$ is the refractive index one should use for computing reflection for TM polarization.

11.8   BEHAVIOR OF $n^{z\prime}$ WITH INCIDENT ANGLE

Similar to n^z, we can examine how $n^{z\prime}$ varies as a function of incidence angle. Figure 11.4 shows the values for $n_f^{z\prime}$ for the same materials as in Figure 11.2. The general behavior in this case is that all the values *increase* from their normal-incidence values. As before, the materials with lower refractive indices experience a larger change with the incidence angle. Additionally, unlike in Figure 11.2, in this case we have intersecting curves. In other words, $n_f^{z\prime}$ can become equal to the incident medium. From equation (11.25), this would lead to a reflection of zero at this particular angle. This is known as Brewster's angle, and it results in zero reflection for TM polarized light at that specific angle.

Since Brewster's condition corresponds to $n_f^{z\prime} = n_a^{z\prime}$, we can use this condition to solve for the Brewster's angle θ_B. This derivation proceeds as follows:

$$\frac{n_f^2}{n_f^z} = \frac{n_a^2}{n_a^z} \tag{11.26}$$

$$\frac{n_f^2}{\sqrt{n_f^2 - n_a^2 \sin^2 \theta_B}} = \frac{n_a}{\cos \theta_B} \tag{11.27}$$

$$\frac{n_f^4}{n_f^2 - n_a^2 \sin^2 \theta_B} = \frac{n_a^2}{\cos^2 \theta_B}. \tag{11.28}$$

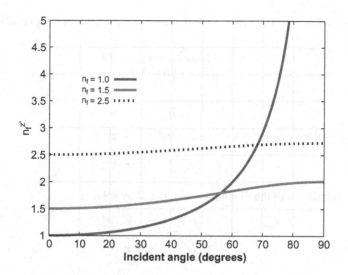

Figure 11.4 Calculated values of $n_f^{z\prime}$ for $n_f = 1.0$, 1.5, and 2.5.

We can use the trigonometric relations

$$\sin^2 \theta_B = \frac{\tan^2 \theta_B}{1 + \tan^2 \theta_B} \tag{11.29}$$

$$\cos^2 \theta_B = \frac{1}{1 + \tan^2 \theta_B} \tag{11.30}$$

to get

$$\frac{n_f^4 \left(1 + \tan^2 \theta_B\right)}{n_f^2 \left(1 + \tan^2 \theta_B\right) - n_a^2 \tan^2 \theta_B} = n_a^2 \left(1 + \tan^2 \theta_B\right), \tag{11.31}$$

which can be finally simplified to

$$\tan \theta_B = \frac{n_f}{n_a}. \tag{11.32}$$

This is the well-known Brewster's angle formula between two materials. From Figure 11.4, we can identify that the intersection of the curves for air and $n_f = 1.5$ occurs at an angle of 56.35°. For the second material with $n_f = 2.5$, Brewster's angle occurs at 68.2°. Both of these values are in agreement with $\tan^{-1}(1.5)$ and $\tan^{-1}(2.5)$ from equation (11.32).

11.9 SINGLE-LAYER ANTIREFLECTION FOR TM INCIDENCE

The single-layer antireflection principles can also be applied for TM polarization. The procedure is very similar to TE, except n^z has to be used to determine the film thickness, while $n^{z\prime}$ has to be used to determine the film index. This is because the phase is determined by n^z for both TE and TM, which can be verified from equations (11.6) and (11.19).

Using $n_s = 1.5$, $\theta = 45°$, and $\lambda_0 = 550$ nm, we can get the following results for TM:

$$n_s^{z\prime} = \frac{n_s^2}{\sqrt{n_s^2 - n_a^2 \sin^2 \theta}} = 1.70 \tag{11.33}$$

$$n_a^{z\prime} = \frac{n_a}{\cos \theta} = 1.41. \tag{11.34}$$

From this, the film's index $n_f^{z\prime}$ can be calculated, which gives

$$n_f^{z\prime} = \sqrt{n_s^{z\prime} n_a^{z\prime}} = 1.55. \tag{11.35}$$

The material index n_f of the film has to be obtained by solving for

$$n_f^{z\prime} = \frac{n_f^2}{\sqrt{n_f^2 - n_a^2 \sin^2 \theta}}. \tag{11.36}$$

This requires a quadratic solution, and we can get

$$n_f^2 = \frac{n_f^{z\prime 2} + \sqrt{n_f^{z\prime 4} - 4 n_f^{z\prime 2} n_a^2 \sin^2 \theta}}{2} \tag{11.37}$$

from which we can get $n_f = 1.30$.

To determine the film thickness, we need to use n_f^z (not $n_f^{z\prime}$), which gives

$$n_f^z = \sqrt{n_f^2 - n_a^2 \sin^2 \theta} = 1.09. \tag{11.38}$$

Therefore, the required film thickness is $t_f = \frac{\lambda_0}{4 n_f^z} = 125.7$ nm. The corresponding reflection spectrum of this structure is shown in Figure 11.5. The behavior of this curve is very similar to the TE antireflection shown in Figure 11.3, except the film index and thicknesses are significantly different for TM.

11.10 EFFECTIVE REFLECTANCE INDEX CONTOURS

Following the principles developed in Chapter 3, we can plot the effective reflectance index contours for oblique incidence. However, the contours will be different for TE and TM. Using equations (11.5) and (11.6), we can write the contour for TE polarization as

$$n_r = n_f^z \frac{\left(n_s + n_f^z\right) + \left(n_s - n_f^z\right) e^{-2jk_0 n_f^z z}}{\left(n_s + n_f^z\right) - \left(n_s - n_f^z\right) e^{-2jk_0 n_f^z z}}. \tag{11.39}$$

Similarly, using equations (11.18) and (11.19), the contour for TM polarization becomes

$$n_r = n_f^{z\prime} \frac{\left(n_s + n_f^{z\prime}\right) + \left(n_s - n_f^{z\prime}\right) e^{-2jk_0 n_f^\prime z}}{\left(n_s + n_f^{z\prime}\right) - \left(n_s - n_f^{z\prime}\right) e^{-2jk_0 n_f^z z}}. \tag{11.40}$$

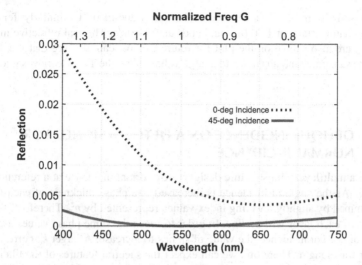

Figure 11.5 Reflection spectrum for TM polarization from a single-layer antireflection layer designed for 45° incidence.

Notice that the phase term contains n_f^z for both polarizations. Hence the film thickness will be determined by n_f^z for both polarizations. The starting and ending points of the TE trace will be determined by n_f^z, and for the TM trace, it will be determined by n_f^y.

The single-layer antireflection condition is depicted in Figure 11.6 for a substrate with a refractive index of $n_s = 1.48$ at 45° incidence. For the TE case, n_s^z is 1.32, and n_a^z is 0.71. Therefore, the contour begins at the equivalent substrate index of 1.32 and makes

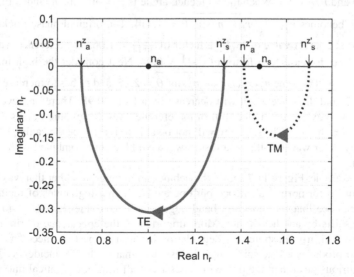

Figure 11.6 TE and TM reflectance index contours for a single-layer antireflection on a substrate with $n_s = 1.48$ and $n_a = 1.0$.

a clockwise half-circle to reach the equivalent air value of 0.71. Similarly, for the TM case, the contour starts at 1.70 ($n_s^{z\prime}$) and ends at 1.41 ($n_a^{z\prime}$). The real refractive indices of n_s and n_a are also shown on the plot for reference. We can also see that the TM case moves the substrate (and air) index to a large value, while the TE case moves it to a lower value.

11.11   OBLIQUE INCIDENCE ON A FILTER DESIGNED FOR NORMAL INCIDENCE

Consider a multilayer film structure designed for normal incidence at a reference wavelength λ_0. As the angle of incidence is increased, the phase thickness of each film will be determined by slightly lowering index values represented by n_f^z. Therefore, the phase thickness of each layer will decline. In order to reach the same phase values as those in the case of normal incidence, the value of k_0 has to increase. A larger k_0 corresponds to a shorter wavelength. Therefore, we can expect the spectral features of the filter to shift toward shorter reference wavelengths. This is the reason for the blue shift observed with increasing incident angles.

Unfortunately, this shift in wavelength is not a simple lateral shift of the entire spectrum. The overall performance of the filter will become distorted in addition to the blue shift. This occurs due to two separate factors: (1) the values of n_f^z will vary by different amounts in each film, which makes it impossible to reach the same phase shifts that were present at normal incidence, and (2) the amplitudes of the reflections will be determined by n_f^z for TE, and by $n_f^{z\prime}$ for TM, which will produce an additional polarization-dependent spectral distortion.

For example, consider a structure with several layers with refractive indices of $n_{f1} = 1.5$ and $n_{f2} = 2.5$. When the incidence angle is $\theta = 30°$, the n_f^z index of the low-index film becomes $n_{f1}^z = \sqrt{n_{f1}^2 - n_a^2 \sin^2\theta} = 1.414$. The original phase thickness will now occur at a larger value of k_0, by a factor of $\frac{1.5}{1.414} = 1.06$. The reference wavelength will therefore decrease by a factor of $\frac{1.414}{1.5} = 0.94$. Next consider the high-index film. It's n_f^z index will become $n_{f2}^z = \sqrt{n_{f2}^2 - n_a^2 \sin^2\theta} = 2.45$, and k_0 has to increase by a factor of 1.02, and the wavelength will decrease by a factor 0.98. Therefore, even though each layer will experience a blue shift of its reference wavelength, the exact shift will be different in each layer. The structure will not be able to reproduce the same phase thickness at any other wavelength. As a result, we see a blue shift combined with a spectral distortion.

As an example, Figure 11.7 shows the behavior of a long-pass filter that was designed and optimized for normal incidence but operated at increasing angles of incidence. At 30°, we can see that the reflection band edge for TE has blue-shifted by 40 nm, and at 45°, it shifts by another 40 nm. More importantly, the appearance of ripples inside the passband clearly illustrates the degradation in the filter performance. The TM incidence also produces a blue shift, but it is less than that of the TE incidence. This may be a bit curious given that the phase thicknesses of all films are identical under TE and TM. The difference in blue shifts arises due to the smaller difference in $n^{z\prime}$ between the films for TM polarization. This results in a reduced reflection bandwidth of the structure,

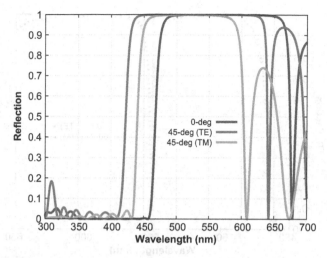

Figure 11.7 Reflection spectrum for the numerically optimized short-pass filter (same as Figure 8.9b in Chapter 8) for normal incidence and 45° incidence.

producing a red shift of the band edge (at the short-wavelength edge of the reflection band). Combined with the blue shift due to the z-phase indices, this results in a smaller overall blue shift for TM compared to TE.

11.12 MULTILAYER FILTERS DESIGNED FOR OBLIQUE INCIDENCE

Multilayer antireflection coatings, or other filters such as edge filters and line filters, can be designed to operate at a specific incident angle. This is done using the same techniques discussed in the previous chapters, except we have to treat TE and TM separately. For TE, we have to use n^z as the refractive index. For TM, we have two refractive indices – n^z for film thicknesses and $n^{z'}$ for amplitudes. These conversions are required only for the equivalent index models. Numerical models such as the transfer matrix method (TMM) do not require the use of these modified indices, although one could use these indices and treat an oblique-incidence problem as a normal-incidence problem.

Consider the $\frac{H}{2}R_{5H}HR_{5H}\frac{H}{2}$ line-pass filter discussed in Figure 9.1 in Chapter 9. We can redesign that filter for operation as 60° incidence as follows. The high and low film indices were 2.5 and 1.5. At 60° incidence, the phase indices n^z_f of these films become 2.345 and 1.225, respectively. To achieve line pass at the same wavelength of 550 nm, the quarter-wave film thicknesses would have to be $\frac{550}{4\times2.345} = 58.63$ nm and $\frac{550}{4\times1.225} = 112.3$ nm. The calculated transmission for this cavity is shown in Figure 11.8. We can see that the transmission line is maintained at 550 nm when the incident angle is 60° for both polarizations. However, the line width for TE is much narrower than that for TM. The reason for this becomes apparent if we calculate $n^{z'}$. These are 2.665 and 1.837. The interface reflections for TE will be determined by n^z, which are 2.345 and 1.225. The TE interface reflections will be $\left(\frac{2.345-1.225}{2.345+1.225}\right)^2 = 9.8\%$ and $\frac{2.665-1.837}{2.665+1.837} = 3.4\%$. As a result, the resonance will be much stronger under TE illumination than under TM illumination, resulting in a much narrower transmission line width.

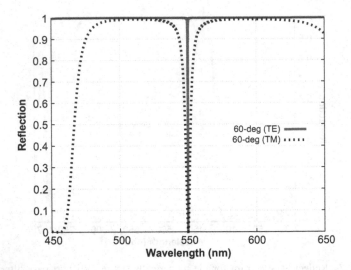

Figure 11.8 Reflection spectrum of the single-cavity line filter $\frac{H}{2}R_{5H}HR_{5H}\frac{H}{2}$ using film indices of 2.5 and 1.5 designed for $60°$ incidence for TE and TM polarizations.

11.13   A COMMON MISCONCEPTION

One common misconception arises from viewing the fields in thin films as rays instead of waves. In this view, a tilted ray will travel a longer path through the film compared to a normally incident ray. A longer path will lead to a large phase angle. Since the phase is k_0nL where L is the distance, this argument will lead one to conclude that k_0 has to get smaller (or wavelength has to get longer) to achieve the same phase as the normally incident ray. In other words, one might conclude that a tilted incidence should produce a red shift in performance. We know from our derivations, and TMM simulations, that that is not correct. The basis of this erroneous argument lays in treating the incident field as a ray instead of a wave. This is illustrated in Figure 11.9.

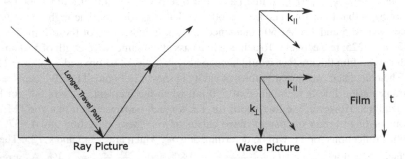

Figure 11.9 Ray versus wave picture to explain the blue spectral shift due to oblique incidence on a thin film.

In the wave treatment, the wave number k inside a film has two orthogonal components. One component is perpendicular to the films, and the other is parallel to the films. If we label these as $k_\parallel$ and $k_\perp$, we can get $k_\parallel = k\sin\theta$ and $k_\perp = k\cos\theta$. The perpendicular field component, $k_\perp$, is the wave component that produces interference between film boundaries. The parallel component, $k_\parallel$, travels parallel to the surfaces and does not contribute to interference. In other words, $k_\perp$ is the component responsible for the spectral behavior. The parallel component $k_\parallel$ plays an important role in the boundary conditions – its value has to be equal across all layers in order to satisfy the field continuity across the film boundaries. Since $k_\perp$ gets smaller with increasing angle (due to the $\cos\theta$), k has to get *larger* in order to achieve a phase same as that of the normal-incidence case. This results in the blue shift in spectral performance, which is the expected result.

11.14 THIN-FILM POLARIZING BEAM SPLITTER

Polarizing beam splitters are typically made from birefringent crystals, but they can also be accomplished using a multilayer thin-film structure. The thin-film construction has the benefit of being adaptable to many different wavelengths as well as being compact. The concept is based on utilizing a high-reflection stack, such as $\left(\frac{H}{2}L\frac{H}{2}\right)^N$ or $\left(\frac{L}{2}H\frac{L}{2}\right)^N$, but in this case, we design it to operate at the Brewster's angle of incidence [3]. Under TE, the refractive index contrast between n_1^z and n_2^z will be enhanced, resulting in a broad reflection band (broader than even that at normal incidence). Under TM, $n_1^{z\prime}$ and $n_2^{z\prime}$ will become equal, resulting in a zero reflection. Therefore, the reflected beam will be TE polarized, and the transmitted beam will be TM polarized. This is the basic concept behind thin-film polarization splitters.

Consider a system where the incident material has an index n_a with an angle of incidence θ, with two alternating thin films of indices n_{f1} and n_{f2} on a substrate with an index of n_s. Brewster's condition between n_{f1} and n_{f2} will occur when the incident angle *inside* film #1 is $\theta_{f1} = \tan^{-1}\left(n_{f2}/n_{f1}\right)$. Using Snell's law, we can relate this angle to the incident angle in the incident medium as

$$n_a \sin\theta = n_{f1}\sin\theta_{f1} \tag{11.41}$$

$$= n_{f1}\sin\left(\tan^{-1}\frac{n_{f2}}{n_{f1}}\right) \tag{11.42}$$

$$= \frac{n_{f1}n_{f2}}{\sqrt{n_{f1}^2 + n_{f2}^2}} \tag{11.43}$$

from which we can get

$$\sin\theta = \frac{n_{f1}n_{f2}}{n_a\sqrt{n_{f1}^2 + n_{f2}^2}}. \tag{11.44}$$

The variables in this expression are n_{f1}, n_{f2}, n_a, and θ. We can calculate any one of these quantities if the other three are known. However, not all combinations will yield a real incident angle or realistic refractive index values. In fact, it is difficult to achieve Brewster's condition between two films when the incident beam is coming from air. However, if the incident medium has a higher index, for example 1.5, it is possible to find realistic indices and angles. Figure 11.10 shows two plots of n_{f2} versus n_{f1} for various incident angles.

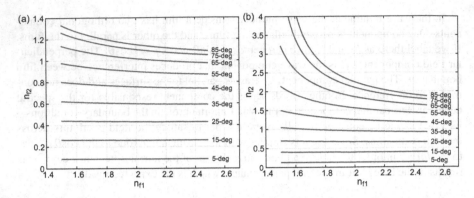

Figure 11.10 Solution of n_{f2} versus n_{f1} from equation (11.44) for two different values of n_a. (a) $n_a = 1.0$, (b) $n_a = 1.5$.

Figure 11.10a is for $n_a = 1.0$, and Figure 11.10b is for $n_a = 1.5$. From Figure 11.10b, we can find that at 45° incidence, $n_{f2} = 1.38$ and $n_{f1} = 1.658$ will satisfy the Brewster's condition. The 45° incidence is useful as a practical consideration because it will allow the TE and TM polarized beams to be separated by 90°. The incident medium index of $n_a = 1.5$ can be achieved by using a silica glass right-angled prism. The films are deposited on the base of a prism, and a second prism is used to sandwich the film structure such that the exit medium also has an index of 1.5. This will separate the TE and TM to be separated by 90°. The configuration is shown in Figure 11.11.

Due to the small refractive index contrast between n_{f1} and n_{f2}, we need to use a greater number of unit cells to achieve the a TE reflection coefficient close to 100%. The layer structure we will use is $\left(\frac{H}{2}L\frac{H}{2}\right)^{15}$. An antireflection coating will have to be

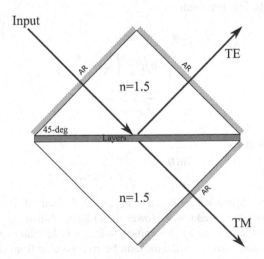

Figure 11.11 Polarizing filter stack sandwiched between two right-angled prisms.

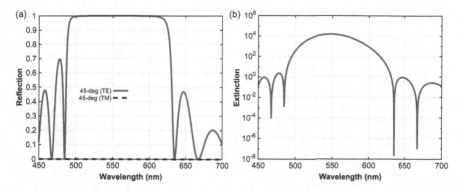

Figure 11.12 Performance of the polarization beam splitting thin-film structure utilizing $\left(\frac{H}{2}L\frac{H}{2}\right)^{15}$ with $n_{f1} = 1.658$ and $n_{f2} = 1.38$. (a) TE and TM reflections at 45° incidence. (b) Extinction ratio $\frac{R_{TE}}{R_{TM}}$.

used on the substrate to allow the TM polarization to pass through without any reflection. Due to the 45° incidence, using $n_{f2} = 1.38$ and $n_{f1} = 1.658$, we can get $n_{f2}^z = 0.8828$ and $n_{f1}^z = 1.2743$. Assuming operation at a reference wavelength of $\lambda_0 = 550$ nm, the quarter-wave film thicknesses will be $\frac{550}{4\times0.8828} = 155.75$ nm and $\frac{550}{4\times1.2743} = 107.90$ nm. Using these values, the reflections of TE and TM transmissions as well as the ratio $\frac{R_{TE}}{R_{TM}}$ are shown in Figure 11.12. The device provides polarization separation over a broad range of wavelengths, in this case from about 500 to 600 nm, with extinction ratios ranging between 100 and 10^5. In commercially available devices, the extinction ratios are on the order of 10^3. Practical performance figures are often lower than calculated values because even very small amounts of scattering of the TM field due to roughness at any of the surfaces can significantly degrade the extinction ratio.

11.15 PROBLEMS

1. Modify the line filter shown in Figure 9.14 in Chapter 9 for 632 nm wavelength and an incident angle of 30°. Examine how the filter behaves for TE and TM polarization.
2. Design a polarizing beam splitter with an extinction ratio of at least 100 using a rutile (TiO_2) right-angle prism and ZnS as one of the film materials, for a reference wavelength of 1,550 nm.

REFERENCES

1. Qi, H., Hong, R., Yi, K., Shao, J. & Fan, Z. Nonpolarizing and polarizing filter design. *Applied Optics* **44**, 2343–2348 (April 2005).
2. Amotchkina, T. V. Empirical expression for the minimum residual reflectance of normal- and oblique-incidence antireflection coatings. *Applied Optics* **47**, 3109–3113 (June 2008).
3. Li, L. & Dobrowolski, J. A. Visible broadband, wide-angle, thin-film multilayer polarizing beam splitter. *Applied Optics* **35**, 2221–2225 (May 1996).

12 Metal Film Optics

12.1 OPTICAL PROPERTIES OF METALS

Thin-film structures with metals or semimetals can serve some important purposes. One is the possibility of electrical conduction while being reasonably transparent to certain wavelengths. ITO (indium tin oxide) is a perfect example of this. The second is the possibility of simpler layer structures for achieving short-pass and band-pass filters, albeit with compromised performance. While the design principles are nearly the same as with any other dielectrics, the complex refractive indices of these materials warrants a separate discussion.

There are several definitions of what constitutes a metal, but in optics, a metal is defined as any material with a negative ε_1. This is the real part of the complex permittivity ε_c, which was defined in equations (1.37) and (1.38) of Chapter 1:

$$\Re\{\varepsilon_c\} = m^2 - \kappa^2 \tag{12.1}$$

$$\Im\{\varepsilon_c\} = 2m\kappa. \tag{12.2}$$

We should be able to verify from the properties of metals listed in Chapter 2 that they all have $\kappa > m$, which would make $\Re\{\varepsilon_c\}$ negative. Exceptions occur at deep-UV wavelengths where $\Re\{\varepsilon_c\}$ becomes positive for many metals, which correspond to dielectric behavior. The $\Re\{\varepsilon_c\}$ of most metals increase (in negative values) with increasing wavelength (stronger metallic behavior at longer wavelengths). On the other hand, semiconductors can exhibit metallic behavior at shorter wavelengths and dielectric behavior at longer wavelengths. Some phase change materials exhibit both characteristics, which will be examined in more detail in Chapter 13. For our purposes, it is not necessary to dwell on the precise definition of metals. We can loosely consider all materials with high electrical conductivity and large values of κ as having metallic characteristics. The only information we really need to accurately model the optical behavior of metals is their complex refractive index values as a function of wavelength.

Metals exhibit much higher reflections than any dielectrics. This arises primarily due to the large κ. The real part can be close to 1.0 or even lower depending on the wavelength and type of metal. Assuming the metal has a low m and a high κ, we can apply the normal-incidence-reflection formula to gain some understanding.

$$r = \frac{n_a - n_s}{n_a + n_s} \approx \frac{+j\kappa}{-j\kappa} \to -1. \tag{12.3}$$

In other words, the reflection from a metal is high primarily due to the large κ value compared to m. For example, silver has a refractive index of $0.125 - j3.3$ at $\lambda = 550$ nm. Using equation (12.3), we can determine that the reflection $|r|^2$ will be 96%. Furthermore, the refractive index of metals can be strongly dependent on wavelength, especially at wavelengths that are close to their plasma frequency. This dispersion is what gives certain metals such as copper and gold their characteristic color.

The reflection spectra of most common optical metals (assuming an infinitely thick metal substrate) is shown in Figure 12.1. In the visible spectrum, silver has the highest reflectivity of nearly 98%, but it quickly falls near its plasma frequency. The reflectivity of silver at the UV wavelength of 325 nm is only about 6%. Also, silver is very soft and often unsuitable as a robust coating. It also tarnishes due to reaction with sulfur in the atmosphere. Aluminum has a broader spectral reflection from UV to IR, but its reflection is lower than silver. It dips below 90% near 800 nm due to an increase in the absorption loss. An important advantage of aluminum is that aluminum oxide makes a very tough and robust passivation. As a result, aluminum is the most used reflector in household and laboratory mirrors. Both copper and gold are unsuitable as reflectors in the visible range, but gold makes an excellent infrared reflector. It is also inert to environmental factors, so it is widely used in infrared equipment. Chromium has a low reflection, but it has excellent adhesion to many surfaces; therefore it is often used as an underlayer for gold, silver, etc.

Metals offer the possibility of making edge, band, and notch filters with far fewer layers compared to dielectric films alone, albeit with some unavoidable absorption losses as well as less-perfect transition characteristics. The electrical conductivity is also an advantage, which allows the possibility of combining optical performance with electrical conductivity. ITO is currently the most widely used transparent conductor, but its conductivity is very modest which limits its high-speed performance.

Following the same principles developed earlier, we can design antireflection coatings for metals. However, unlike in dielectrics, antireflection here does not

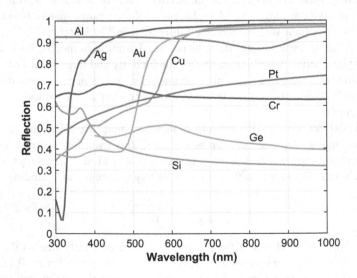

Figure 12.1   Reflectivity of various bulk metals and semiconductors.

always correlate to a maximum transmission due to the presence of absorption. Therefore, antireflection in metals falls under two different categories:

1. antireflection to reduce reflection and increase absorption,
2. antireflection to reduce reflection and maximize transmission.

Each of these aspects will be discussed separately.

12.2 TRANSPARENCY OF METALS

Absorption in a metal film is primarily determined by its loss tangent. This was derived in Chapter 1, Section 1.4, as

$$\tan \delta = \frac{2m\kappa}{|m^2 - \kappa^2|}. \tag{12.4}$$

The loss tangents of several metals, semiconductors, and ITO are shown in Figure 12.2. We can see that silver has the lowest loss tangent of all metals in the visible spectrum, followed by copper and gold. However, ITO and silicon have even lower values. In fact, dielectric materials like SiO_2 would have even lower values (actually, pretty close to zero). Therefore, the loss tangent value by itself is not the most meaningful comparison. We need to consider both the loss tangent and the electrical conductivity simultaneously.

For this reason, we will define the product $\rho \tan \delta$ as a combined indicator of conductivity and transparency. The DC resistivities of these metals are listed in Table 12.1. The calculated values of $\rho \tan \delta$ for the same materials from Figure 12.2 are shown in Figure 12.3. With this definition, silver comes out with the lowest value

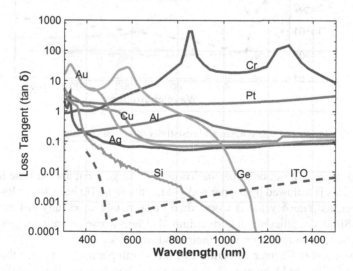

Figure 12.2 Loss tangents of various metals and semiconductors.

Table 12.1
Resistivity Values of Materials Shown in Figure 12.2

Metal	ρ (nΩ·m)
Ag	15.9
Au	24.4
Cu	16.8
Cr	138
Al	26.5
Pt	106
ITO[a]	5,000
Si[b]	5×10^8
Ge[b]	1×10^9

[a] Resistivity of ITO is highly dependent on deposition conditions.
[b] Resistivity values of Si and Ge are highly dependent on impurity concentrations.

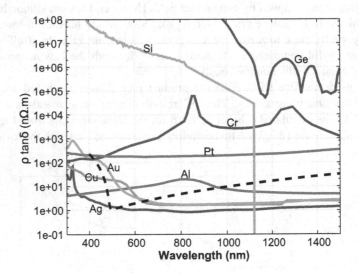

Figure 12.3 Product of loss tangent and resistivity of various metals.

of $\rho \tan \delta$ because it not only has the lowest loss tangent but also has the lowest resistivity. This is followed by copper and gold. Although ITO has a lower loss tangent than silver, its $\rho \tan \delta$ value is worse than most metals, especially at longer wavelengths. Similarly, silicon and germanium also fare worse when their loss tangents are compared in combination with their conductivities.

As discussed in Chapter 1 Section 1.4, it is worth pointing out again that the loss tangents of silver, gold, copper, etc. are small not because they have small values of κ but because they have small values of m compared to κ. As a result, κ alone is not indicative of absorption loss. It is a common misconception that a large κ implies more absorption. In fact, κ is simply the field attenuation factor. The absorption loss will be determined by the loss tangent, which includes both m and κ.

12.3 ANTIREFLECTION DESIGNS FOR METAL SUBSTRATES

Though it may seem counterintuitive, it is possible to make antireflection for metal substrates. This works on the same principles of dielectric substrates, which were discussed in Chapters 3 and 5. Layers should be designed on top of the metal such that the final effective reflectance index n_r approaches that of the outside medium (air). This allows all of the incident energy to be transmitted into the metal substrate. However, the high reflection of the substrate necessitates the placement of a second reflector to produce an equally high reflection with the opposite phase. This inevitably results in a narrow antireflection band. Therefore, these structures will behave more like line filters than broadband antireflectors. Additionally, unlike the case with dielectric substrates, the transmitted energy will be entirely absorbed by the metal unless the metal substrate has partial transparency. Remember that n_r is used for predicting reflection, not transmission. The effective reflectance index n_r discussed so far has been on dielectric substrates with real refractive indices. Even though we examined substrates with a complex n_r due to existing films on the substrate (Chapter 5, Section 5.6), all of the materials in the system still had real refractive indices. In this section, we will consider substrates and films that have complex material indices.

Let's consider silver as the substrate. At $\lambda_0 = 550$ nm, silver has a refractive index of $0.125 - j3.35$. This will place the starting point of the effective reflectance index contour on the lower left portion of the complex plane (using $n_a = 1.0$ as the center). To find a single film that will produce antireflection, we can draw contours for several film indices that start from $n_r = 1.0$ to see if any of them come close to intercepting the silver refractive index of $0.125 - j3.35$. This is shown in Figure 12.4. We can see that none of the films intercept the refractive index value of silver. Therefore, we can conclude that it is not possible to produce antireflection on a silver substrate with a single dielectric layer. Physically this can be understood as arising due to the significantly lower reflection from the dielectric–air interface compared to the metal–dielectric interface. Considering a very large value of $n_f = 4.0$, the reflection at the dielectric–air interface will be $\left|\frac{4-1}{4+1}\right|^2 = 0.36$, and at the metal–dielectric interface, it will be $\left|\frac{4-(0.125-j3.35)}{4+(0.125-j3.35)}\right|^2 = 0.93$. This large difference in the reflection amplitude will lead to an incomplete destructive interference regardless of the phase thickness of the dielectric film.

12.3.1 USING FILMS WITH COMPLEX REFRACTIVE INDICES

Even though it is not possible to find a single dielectric film that can cancel the reflection from a silver substrate, we can find a single-layer antireflection solution if we allow the film to have a complex refractive index. This actually results in a

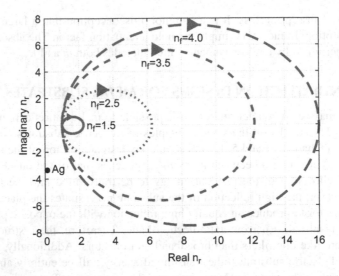

Figure 12.4 Contour of different film indices starting from 1.0. The location of silver's refractive index is at the bottom left corner.

continuum of solutions rather than a discrete solution. We will approach this problem by setting the real part of the refractive index of the film and solving for the required imaginary part. The effective reflectance index was defined in Chapter 3 equation (3.16) as

$$n_r = n_f \frac{(n_s + n_f) + (n_s - n_f) e^{-j2\theta}}{(n_s + n_f) - (n_s - n_f) e^{-j2\theta}},$$ (12.5)

where n_s and n_f are the refractive indexes of the metal substrate and the film, respectively. In this case, we will separate the film's index into its real and imaginary parts, such as $m_f - j\kappa_f$, where it is implied that m_f and κ_f are both real. The phase thickness will also be complex. However, the real and imaginary parts of the phase are not independent; they will scale at the same ratio as the real and imaginary parts of the refractive index. As a result, we can write the complex phase as $\theta\left(1 - j\frac{\kappa_f}{m_f}\right)$ where θ is the real part of the phase. This allows us to rewrite equation (12.5) as

$$n_r = n_f \frac{(n_s + n_f) + (n_s - n_f) e^{-j2\theta\left(1-j\frac{\kappa_f}{m_f}\right)}}{(n_s + n_f) - (n_s - n_f) e^{-j2\theta\left(1-j\frac{\kappa_f}{m_f}\right)}}.$$ (12.6)

To achieve antireflection, equation (12.6) has to be equal to 1.0. If we consider m_f to be fixed, and n_s is known, we will have two unknowns, κ_f and θ, but only one equation. However, it should be noted that equation (12.6) is complex, and therefore it really contains two equations. Therefore, the required condition can be written as

$$\Re\{\text{Eqn (12.6)} - 1\} = 0, \tag{12.7}$$

$$\Im\{\text{Eqn (12.6)}\} = 0. \tag{12.8}$$

Given m_f and n_s, we can now solve for κ_f and θ. Figure 12.5 shows the solution of κ and θ as a function of m_f assuming the refractive index of the silver substrate to be $0.125 - j3.35$ (at 550 nm). The figure illustrates that one can find many combinations of n_f and κ that will meet the antireflection condition for silver (or any other metal, for that matter). Figure 12.6 shows the contours corresponding to these solutions, where we can verify that all of them begin at the refractive index of silver and terminate at $n_r = 1.0$. Each refractive index takes a slightly different contour, but the starting and ending points are exactly the same.

Using one of these solutions of $2.0 - j0.575$ with a phase thickness of $0.758\frac{\pi}{2}$ (which corresponds to a physical thickness of 52 nm at a reference wavelength of 550 nm), we can plot the spectral reflectance using the transfer matrix method (TMM). This is shown in Figure 12.7. It clearly confirms that zero reflection is obtained at the reference wavelength of 550 nm.

It is insightful to examine the behavior of contours traced by a film with a complex refractive index. With real dielectrics, we learned that the contours traced a circular arc in the complex plane moving clockwise toward the upper right or toward the lower left, depending on whether the film index was larger or smaller than the substrate index. For phase thicknesses exceeding π, the contours retraced the same circular path over and over again. A film with a complex refractive index has a very different behavior. The real part of the refractive index still forces the contour to trace a circular arc, but the imaginary part forces the trajectory toward the film's refractive

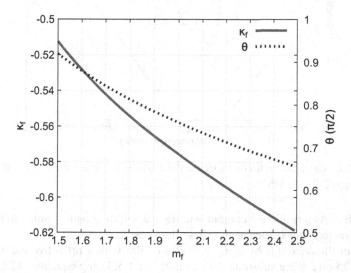

Figure 12.5 Solution of κ and θ (real part) as a function of n_f (real part) of the film index for achieving antireflection condition on a silver substrate.

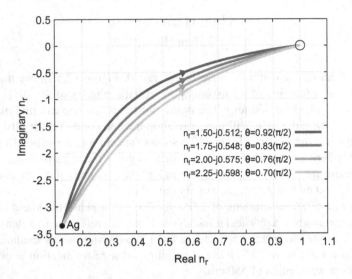

Figure 12.6 Contours of different film indices starting from the silver substrate and ending at $n_r = 1.0$. These contours are the same solutions shown in Figure 12.5.

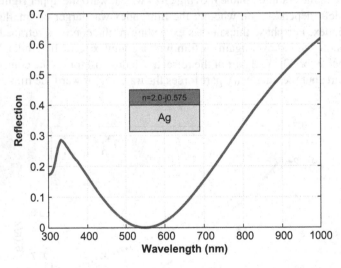

Figure 12.7 Calculated reflection spectrum of a silver substrate with a 52 nm film with an index of $n_f = 2.0 - j0.575$.

index value. As a result, the contour will trace a spiraling path, coming to rest at the film's refractive index value as the phase thickness increases.

We can illustrate this by using a fictitious film with a refractive index of $n_f = 2.5 - j0.25$ on a silica substrate with an index of 1.5. Using equation (12.6), we can plot the resulting contour for increasing values of film thicknesses. This is shown in Figure 12.8. The plot shows the effect of increasing phase thickness θ (real part)

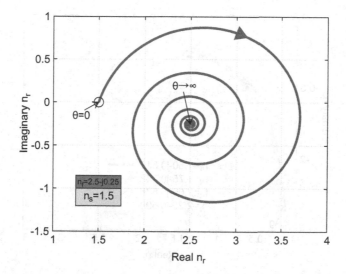

Figure 12.8 Effective index contour of a fictitious film with a complex index of $2.5 - j0.25$ on a substrate with an index of 1.5.

up to ∞. It demonstrates that the contour makes a spiraling path toward the refractive index value of the film as the thickness gets larger. The circular paths are due to the real part of the refractive index, and the convergence toward the final value is due to the imaginary part. A smaller imaginary part compared to the real part will make the spiraling rate slower. A larger imaginary part will make it spiral quicker toward the convergence value. This effect arises due to the optical isolation effect of materials with complex indices. Any material with an imaginary part, such as a metal or semiconductor, when sufficiently thick, will prevent light from penetrating past that layer to reach the underlying films. Therefore, regardless of what is underneath that thick film, the effective reflectance index of the structure will be that of the film alone and will not be influenced by what lays beneath. Additionally, from equation (12.6), we should be able to verify that when θ is complex, $n_r \rightarrow n_f$ as $z \rightarrow \infty$. This means that, regardless of the starting point of the contour, the ending value will always be n_f for complex-valued films.

For the solutions found in Figure 12.6, if we extend the film thicknesses to very large values, their contours will spiral toward their refractive index values. This is shown in Figure 12.9. We can see that all of the films start at the refractive index of silver (at the bottom left), cross the antireflection condition at $n_r = 1.0$, and then spiral toward their own refractive index values.

Finally, it is worth pointing out that even though we found combinations of n_f and κ_f to produce antireflection on metal surfaces, these specific combinations of n_f and κ_f may not exactly correspond to a real thin-film material. There is no guarantee that such a material exists in its natural state. Nevertheless, it is still a physically valid solution and is a useful technique in our tool box. There are cases where such a solution can be used to achieve antireflection.

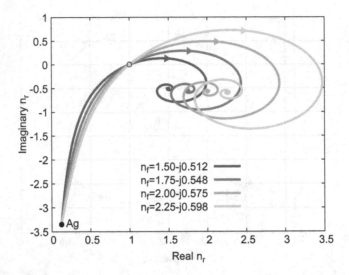

Figure 12.9 When the solutions found in Figure 12.6 are extended to very large thicknesses, the contours converge toward the refractive index values of their respective films.

12.3.2 USING FILMS WITH REAL REFRACTIVE INDICES

As discussed in Figure 12.4, a single film with a real refractive index cannot be used to cancel the reflection from a silver substrate (or most other metals). But it is possible to use multiple layers of dielectrics. However, unlike the multilayer antireflection designs investigated in Chapter 5, because the substrate reflection in this case is very high (about 95% at 550 nm in the case of silver), we would need many dielectric layers. Therefore, the design will be more similar to the line filters discussed in Chapter 9 than to the two- and three-layer designs discussed in Chapter 5.

From the Herpin's equivalence model (Chapter 7), we know that a unit cell such as $\left(\frac{H}{2}L\frac{H}{2}\right)$ will have an equivalent index of jn_1 where n_1 is the outer layer index, and a phase thickness of $\cos^{-1}\left[-\frac{1}{2}\left(\frac{n_1}{n_2} + \frac{n_2}{n_1}\right)\right]$. The goal is to utilize this unit cell along with a dielectric spacer to achieve a final n_r value equal to 1.0. We will assume the dielectric layer has an index 1.5 and thickness t, followed by four unit cells of $\left(\frac{H}{2}L\frac{H}{2}\right)$. We will also assume that $n_1 = 2.5$. We can solve for the phase thickness θ of the dielectric layer, and the value for n_2, such that the contour intersects the real axis at $n_r = 1.0$. Therefore, there are two unknowns (n_2 and θ) and two equations (using the real and imaginary parts of the same contour equation). Following the same approach as described in Section 12.3.1, we can find this solution to be $n_2 = 1.55$ and $\theta = 0.085\left(\frac{\pi}{2}\right)$. The contour plot of this solution is shown in Figure 12.10. The dielectric spacer starts at the silver substrate and converts the effective reflectance index to a value of $0.075 - j2.42$. From there, each unit cell of $\left(\frac{H}{2}L\frac{H}{2}\right)$ progressively moves to the right, coming to the final value of $n_r = 1.0$ after four unit cells.

Figure 12.11 shows the reflection spectrum of this structure calculated using TMM. We can see that it is consistent with the line filters we discussed in Chapter 9.

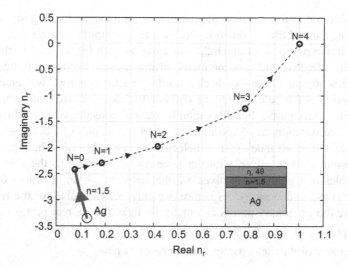

Figure 12.10 Contour plot of the numerical solution of a dielectric layer (refractive index of 1.5 and phase thickness of $\theta = 0.085 \left(\frac{\pi}{2}\right)$) and four unit cells of $\left(\frac{H}{2} L \frac{H}{2}\right)$ with $n_1 = 2.5$ and $n_1 = 1.55$.

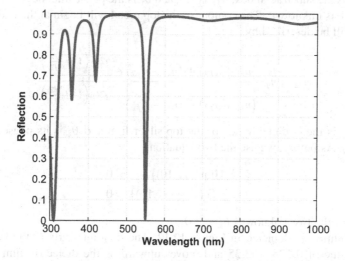

Figure 12.11 Reflection spectrum of the multilayer dielectric structure on a silver substrate shown in Figure 12.10.

12.3.3 ANTIREFLECTION USING METAL–INSULATOR–METAL STRUCTURES

While it is possible to produce resonance between the metal substrate and a multilayer dielectric stack, the simplest approach to creating antireflection on a metal substrate is by using a second metal film to take the place of the top reflector.

Compared to the multilayer line filter structure demonstrated in Figures 12.10 and 12.11, using a metal film as the top reflector greatly simplifies the design.

The solution proceeds in much the same way as with Figure 12.10 - the contour starts at the refractive index of the metal substrate, continues through the dielectric layer, and the top metal brings it back toward $n_r = 1.0$. As before, we can only solve for two variables (since the contour's end point consists of two equations from the real and imaginary parts). For this example, we will consider silver as the substrate, at a reference wavelength of 550 nm, with a refractive index of $0.125 - j3.35$. We will set the refractive index of the dielectric spacer to be 1.5. The top metal will also be assumed to be silver, with the same refractive index as the substrate. The two variables that need to be solved are the dielectric phase thickness θ_1 and the top metal phase thickness θ_2. As before, we only need to solve for the real part of θ_2 because the imaginary part is automatically linked to its real part, such that the complex phase is $\theta_2 \left(1 - j \frac{\kappa_{f2}}{m_{f2}} \right)$.

The contour of the dielectric spacer can be written as

$$n_{r1} = n_{f1} \frac{(n_s + n_{f1}) + (n_s - n_{f1}) e^{-j2\theta_1}}{(n_s + n_{f1}) - (n_s - n_{f1}) e^{-j2\theta_1}}, \tag{12.9}$$

where n_s is the substrate index, n_{f1} is 1.5, and θ_1 is the phase thickness of the dielectric (which is unknown). The contour of the top metal will begin at the ending value n_{r1} and will be described by

$$n_{r2} = n_{f2} \frac{(n_{r1} + n_{f2}) + (n_{r1} - n_{f2}) e^{-j2\theta_2 \left(1 - j \frac{\kappa_{f2}}{m_{f2}} \right)}}{(n_{r1} + n_{f2}) - (n_{r1} - n_{f2}) e^{-j2\theta_2 \left(1 - j \frac{\kappa_{f2}}{m_{f2}} \right)}}, \tag{12.10}$$

where n_{f2} is the complex index of the top silver film and θ_2 is its phase thickness (real part). As before, we use the two equations

$$\Re \{\text{Eqn (12.10)}\} - 1 = 0 \tag{12.11}$$
$$\Im \{\text{Eqn (12.10)}\} = 0 \tag{12.12}$$

to solve for the two unknowns θ_1 and θ_2.

The solution is depicted in Figure 12.12. The contour starts at the bottom left from a value of $0.125 - j3.35$ and moves upward as the dielectric film thickness increases. When the phase thickness of the dielectric is $\theta_1 = 1.43 \left(\frac{\pi}{2} \right)$ (corresponding to a physical thickness of 131.2 nm), the first layer ends, and the second silver film begins. As demonstrated in Figure 12.12, the trace asymptotically moves toward $0.125 - j3.35$. Inevitably, the contour has to cross the real axis, and the solution for θ_2 ensures that the crossing coincides with $n_{r2} = 1$. The phase thickness of the metal film at this point is $\theta_2 = 0.03 \frac{\pi}{2}$, which corresponds to a film thickness of 33.6 nm.

We can verify this calculation by using the TMM. This is shown in Figure 12.13. It confirms that the reflection drops to zero at the reference wavelength of 550 nm. Additionally, Figure 12.13 also shows the reflection spectrum using a fixed refractive

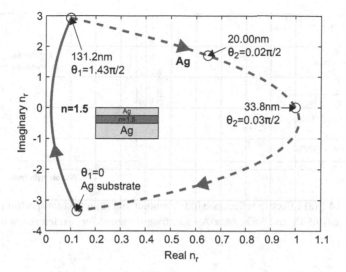

Figure 12.12 Effective reflectance index contour of the two-layer stack on a silver substrate. The first film is a dielectric with an index of 1.5, and the second film is silver.

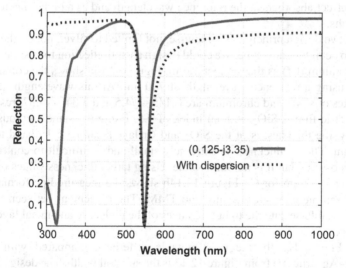

Figure 12.13 Reflection spectrum of the metal–dielectric–metal design shown in Figure 12.13 using a fixed refractive index and a realistic dispersion model.

index of $0.125 - j3.35$ and using a more realistic dispersion model for silver. As expected, the results agree at 550 nm but deviate at other wavelengths. More specifically, since the imaginary part κ of metals generally increase with increasing wavelength, the reflection will also increase as compared to a dispersion-less model.

The narrow spectral band in Figure 12.13 is reminiscent of line-pass filters discussed in Chapter 9. This is primarily due to the large reflection from the top

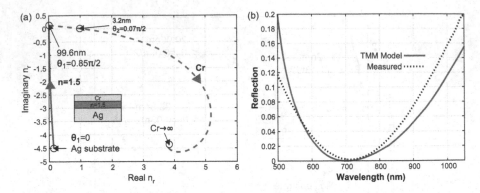

Figure 12.14 (a) Effective reflectance index contour and (b) reflection spectrum (measured and predicted) of Cr and SiO_2 on a Ag substrate designed for a reference wavelength of 700 nm.

and bottom metal interfaces which creates a strong resonant cavity, similar to the dielectric resonant cavities examined in Chapter 9. These structures could be used, for example, in preventing laser reflection from metal surfaces. They could also be used as perfect absorbers at the reference wavelength and as a high reflector at other wavelengths.

The above design principle is, of course, not limited to silver, nor do the two metals have to be of the same type. We could design an antireflection for a silver substrate using chromium (Cr) as the top metal film. Figure 12.14a shows the contour of this structure using a reference wavelength of 700 nm. At this wavelength, the refractive indexes of silver and chromium are $0.145 - j4.5$ and $3.84 - j4.37$, respectively. The dielectric film is SiO_2 with an index of 1.5. Using the same methods outlined previously, the thicknesses of the SiO_2 and Cr layers can be calculated to be 78.2 and 2.5 nm. The Cr thickness is extremely small and is difficult to control exactly during deposition, but it is not impossible. Using target thickness values of 78.2 nm for SiO_2 and 2.5 nm for Cr, Figure 12.14b shows the measured performance from an actual sample compared against the TMM. The agreement between the two is quite remarkable despite the tight tolerances of the dielectric and metal layers during deposition.

From Figure 12.14b, the most notable difference compared with the Ag–dielectric–Ag structure from Figure 12.13 is the spectral width. The design with Cr is significantly wider than that with Ag. This is primarily due to the higher absorption of Cr compared to Ag, resulting in a weaker cavity resonance.

12.4   ANTIREFLECTION ON SEMICONDUCTORS

Semiconductors play an important role in photodetectors, lasers, and integrated optical waveguides. Commonly used semiconductors such as silicon (Si), germanium (Ge), gallium arsenide (GaAs), and indium phosphide (InP) have relatively high refractive indices which can produce as much as 40% reflection if no coating is applied.

When operated above their electronic bandgap (i.e., at longer wavelengths than their bandgap wavelength), the refractive indices will generally be real-valued. Hence the material can be treated just like any other dielectric. However, some applications, such as photodetectors, require operation below the bandgap. In this case, the index will be complex-valued. The material can also become complex-valued at very long wavelengths due to other phenomena such as phonon absorption. In such cases, we can use the same principles discussed in this chapter to design antireflection coatings for semiconductor substrates.

Silicon is the most widely used semiconductor material, so we will begin our discussion with that. A silicon substrate is opaque to visible wavelengths, but the imaginary part of its refractive index is quite small. Its refractive index at a wavelength of 550 nm is $4.08 - j0.04$. Despite this low κ, the reason silicon is opaque has to do with substrate thickness. While thin films of silicon may be reasonably transparent, anything more than a few tens of microns will be opaque. Substrates are typically hundreds of microns thick; hence they are totally opaque to visible light.

On a complex reflectance index plot, silicon's refractive index of $4.08 - j0.04$ places the starting point practically on the real axis because the imaginary part is so small compared to the real part. This allows us to treat it almost like a dielectric material for the purpose of calculating the antireflection layers. In this case, the required film index will be $\sqrt{4.08} = 2.02$, and the required thickness will be close to the quarter-wave thickness of $\frac{\lambda_0}{4n} = 67.7$ nm. The complex index contour traces nearly a semicircle, intersecting at 1.0, as shown in Figure 12.15a. We can, however, use the same method as described in Chapter 5, Section 5.6. Using this approach, we can calculate the exact thickness, which is actually slightly smaller than the quarter-wave thickness. Silicon nitride whose refractive index is close to 2.0 is often used as the antireflection layer for silicon. The resulting reflection spectrum is shown in Figure 12.15b.

The situation is significantly different for germanium, whose refractive index at 550 nm is $5.17 - j2.2$. The imaginary part in this case is not negligible. Using the

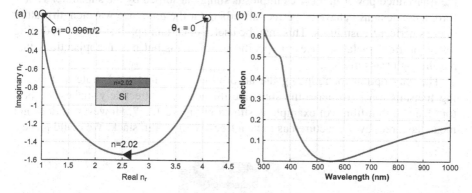

Figure 12.15 Effective reflectance index contour and reflection spectrum of a dielectric layer with an index of 2.02 on a Si substrate. (a) Contour plot for antireflection on silicon. (b) Reflection spectrum.

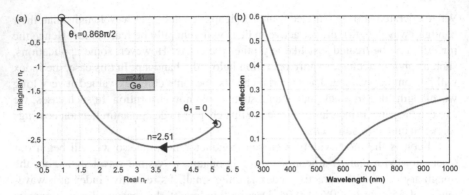

Figure 12.16 Effective reflectance index contour and reflection spectrum of a dielectric layer with an index of 2.503 on a Ge substrate. (a) Contour plot for antireflection on germanium. (b) Reflection spectrum.

same approach discussed in Chapter 5, Section 5.6, we can get the required film index to be $n_f = 2.51$ and the phase thickness to be $\theta = 0.87 \left(\frac{\pi}{2}\right)$. This phase thickness is equivalent to a physical thickness of 47.5 nm.

Figure 12.16a shows the contour of the effective reflectance index, and Figure 12.16b shows the reflection spectrum for the dispersion-less substrate as well as the real germanium substrate. The refractive index of 2.51 is ideally matched to TiO_2, so it can be effectively used as the antireflection film for germanium. The relatively large imaginary value of Ge makes the arc much smaller than a quarter wave.

12.5 BANDPASS FILTERS USING METAL FILMS

So far, we have examined antireflection structures on metal substrates and on semiconductor substrates. But unlike the antireflection designs on dielectric substrates, the transmitted power in these examples is simply absorbed by the substrate. Therefore, these designs behave as perfect absorbers (at the reference wavelength) rather than as perfect transmitters. This may be useful in certain applications for reducing reflection from metal surfaces at specific wavelengths but not in applications that require high transmission.

The only option for reducing the absorption in the metal substrate and allowing transmission is to make the substrate thinner. This effectively makes the bottom metal a thin film. An example is shown in Figure 12.17. However, it does not mean the three-layer structure has to be a free-standing film stack. We could place a

Figure 12.17 $M_1 Q M_2$ single-cavity structure.

dielectric substrate under this three-layer structure, and as long as it is appropriately antireflection coated for $n_r = 1.0$ at the reference wavelength, it can be treated the same as air.

Using this layer structure, it is possible to design an antireflection configuration that exhibits high transmission at the reference wavelength. This is the basic approach that we will explore in this section. However, unlike all-dielectric designs, this design will obviously not exhibit 100% transmission because there will inevitably be some absorption in the metal films.

12.5.1 SINGLE-CAVITY METAL–DIELECTRIC–METAL BANDPASS FILTER

To produce antireflection at the reference wavelength, we need to ensure that the three-layer structure $M_1 Q M_2$ as shown in Figure 12.17 produces a final effective reflectance index n_r equal to 1.0. Due to the complex refractive indices of the metal films, none of the layers can be predefined with a fractional phase thickness such as that of a quarter wave or one-eighth wave as in the case with dielectrics. The goal, nevertheless, is still to make $n_r = 1.0$. In other words, the entire structure must become an absentee structure. This can be done by using the same approach discussed in Section 12.3. To simplify the analysis, we will assume that the cavity is surrounded by symmetric outer regions $n_s = n_a = 1.0$. For $n_s \neq n_a$, we can add layers to the substrate to compensate for the mismatch.

Referring to Figure 12.17, for the bottom metal film M_1, we can write

$$n_{r1} = n_{f1} \frac{(n_s + n_{f1}) + (n_s - n_{f1}) e^{-j2\theta_1 \left(1 + j\frac{\kappa_{f1}}{m_{f1}}\right)}}{(n_s + n_{f1}) - (n_s - n_{f1}) e^{-j2\theta_1 \left(1 + j\frac{\kappa_{f1}}{m_{f1}}\right)}} \tag{12.13}$$

with a starting point of $n_s = 1.0$. For the dielectric film, the effective reflectance index starts at n_{r1} and ends at n_{r2} given by

$$n_{r2} = n_{f2} \frac{(n_{r1} + n_{f2}) + (n_{r1} - n_{f2}) e^{-j2\theta_2}}{(n_{r1} + n_{f2}) - (n_{r1} - n_{f2}) e^{-j2\theta_2}}. \tag{12.14}$$

For the top metal film M_2, the effective reflectance index starts at n_{r2} and ends at n_{r3}. If M_1 and M_2 are the same metals, we can set $n_{f1} = n_{f3}$ (but θ_1 and θ_3 will be different):

$$n_{r3} = n_{f3} \frac{(n_s + n_{f3}) + (n_s - n_{f3}) e^{-j2\theta_3 \left(1 + j\frac{\kappa_{f3}}{m_{f3}}\right)}}{(n_s + n_{f3}) - (n_s - n_{f3}) e^{-j2\theta_3 \left(1 + j\frac{\kappa_{f3}}{m_{f3}}\right)}}. \tag{12.15}$$

In order to create antireflection at the reference wavelength and produce high transmission, the condition that needs to be met is the same as before:

$$n_{r3} = 1.0. \tag{12.16}$$

If the refractive indices of the metal layers and the dielectric are known, the only variables that we need to solve for are θ_1, θ_2, and θ_3. Because there are three independent variables and only two equations (which come from treating the real and imaginary parts of equation (12.15) separately), this system will not have a single unique solution. The approach to finding the solution is to first assume a value for θ_1 and then solve for θ_2 and θ_3. In other words, we have to assume a thickness for the bottom (or top) metal film and then solve for the thicknesses of the other two films.

Let's assume silver as both the top and bottom metals with $n_{f1} = n_{f3} = 0.125 - j3.35$ at a wavelength of 550 nm and a dielectric index of 1.5. First, we need to assume a thickness for M_1. Let's make this to be 15 nm, which makes $\theta_1 = 0.013\frac{\pi}{2}$. Then, numerically solving equation (12.16), we can find $\theta_2 = 1.15\frac{\pi}{2}$ and $\theta_3 = 0.012\frac{\pi}{2}$. These correspond to film thicknesses of 105.9 and 12.7 nm for the SiO_2 and Ag layers, respectively.

Figure 12.18 shows this solution. The contour of the first metal film starts at $n_s = 1.0$ and terminates at $n_{r1} = 0.83 - j1.86$. The dielectric contour starts from this point and runs upward up to $n_{r2} = 0.68 + j1.60$. The contour for M_2 starts from n_{r2} and brings it back to the starting point. Because the starting and ending points of the contours are the same, we can conclude that this behaves as an absentee structure.

Figure 12.19 shows the reflection, transmission, and absorption spectra of this three-layer structure computed using the TMM. Figure 12.19a was obtained using a fixed refractive index of $0.125 - j3.35$ for silver, and Figure 12.19b was obtained using the actual refractive index data for silver with dispersion. The most important aspect to note is the transmission. The peak transmission at the reference wavelength is 75%. At longer wavelengths, the transmission (and reflection) remain relatively flat if the refractive index were to be constant. With a realistic dispersion model

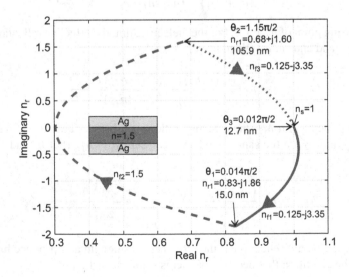

Figure 12.18   Contour of the Ag–dielectric–Ag resonator.

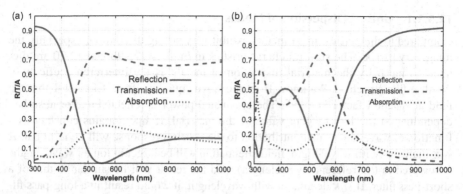

Figure 12.19 Reflection, transmission, and absorption spectra for the Ag–dielectric–Ag structure. (a) Using a fixed refractive index of $0.125 - j3.35$ for silver. (b) Using dispersive refractive index data for silver.

for silver, the spectral plot shows a significant decrease in transmission at longer wavelengths and an increase in transmission at shorter wavelengths. This is because the imaginary part of the refractive index increases rapidly with wavelength reducing the penetration of the field through the silver films.

Figure 12.20 shows the transmission through a single 27.7 nm silver film whose thickness is the same as the combined thickness of the two silver layers in Figure 12.18. The transmission at the reference wavelength is only 15% (compared with 75% in the cavity configuration). As a result, we can conclude that the resonant-cavity configuration significantly increases the forward transmission through the structure compared to the same metal thickness by itself.

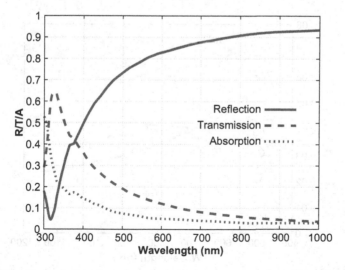

Figure 12.20 Transmission through a single 27.7 nm free-standing silver film (same total thickness as in Figures 12.18 and 12.19.

12.5.1.1 Optical Dispersion of Metals

Compared to dielectrics, most metals exhibit very strong dispersion, especially the imaginary part κ. The spectral characteristics in Figures 12.19b and 12.20 are dictated as much by the material dispersion of metals as the interference effects of the thin-film structure. We know from the wave equations from Chapter 1 that the field attenuation factor is $e^{-k_0\kappa}$. The attenuated power is reflected and/or absorbed depending on the loss tangent value of the material. If $k_0\kappa$ remains constant as a function of wavelength, its contribution to the spectral response will be minimal. If $k_0\kappa$ increases with wavelength, field attenuation will be larger at longer wavelengths (compared to the reference wavelength), and the result will be similar to that of a short-pass filter. If $k_0\kappa$ decreases with wavelength, it would result in a long-pass filter. Most metals of practical interest in optics (i.e., those with reasonably low loss tangents) exhibit an increasing value of $k_0\kappa$ with wavelength. This can be verified from the dispersion characteristics of metals shown in Section 2.4 of Chapter 2. The value of $k_0\kappa$ is plotted in Figure 12.21 for several metals. We can see that silver, gold, copper, and aluminum exhibit $k_0\kappa$ values that increase with wavelength. This will generally result in a short-pass spectral behavior. On the other hand, platinum (not shown) has declining values of $k_0\kappa$, and this will result in long-pass behavior. Chromium has a somewhat flat $k_0\kappa$, and this will lead to a flat response, resulting in a broad antireflection performance. This is partly the reason for the flatter response demonstrated in Figure 12.14 for the antireflection design.

12.5.1.2 Metal–Dielectric–Metal Cavity Structure Layer Thicknesses

For the example in Figure 12.18, we fixed the bottom metal thickness to 15 nm and solved for the other two layers. This calculation can be repeated for other values

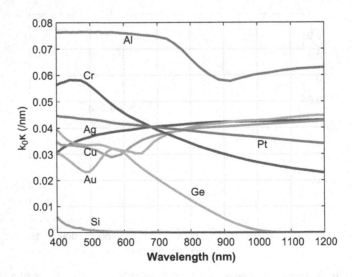

Figure 12.21 $k_0\kappa$ values of various metals.

of bottom metal thicknesses. Figure 12.22 shows this solution. The top metal thickness M_2 and the dielectric thickness Q are shown as a function of the bottom metal M_1 thickness. There are several aspects that are worth noting in this figure. As M_1 thickness increases, M_2 initially increases but then levels off near 34 nm. This can be understood in the context of antireflection designs we discussed for a silver substrate, in Section 12.3. When the bottom metal M_1 becomes thicker, it starts to behave as a bulk silver substrate and isolates any material that lays below. In Figure 12.18, the calculated thicknesses of the dielectric layer and the metal layer are 131.2 and 33.8 nm, respectively, which are indeed the values where the two curves level off in Figure 12.22. Furthermore, we can see that the curve for M_2 does not intersect the $M_1 = M_2$ line. In other words, both the top and bottom metal films cannot be identical in a perfectly resonant structure.

Increasing both M_1 and M_2 will result in increased absorption losses. This is shown in Figure 12.23. The first curve is for $M_1 = 15$ nm, $Q = 105.9$ nm, and $M_2 = 12.7$ nm (total silver thickness of 27.7 nm), and the second curve is for $M_1 = 30$ nm, $Q = 123.4$ nm, and $M_2 = 22.5$ nm (total silver thickness of 52.5 nm). Clearly, the structure with more metal exhibits a lower transmission, although the resonance condition is still maintained at the reference wavelength. Furthermore, both curves will have zero reflection at the reference wavelength even when the metal thicknesses are different. In other words, only the transmission is affected by the total metal thickness. The most noticeable difference between the two structures is the width of the transmission line. As expected, the structure with thicker metal layers exhibits a narrower transmission width.

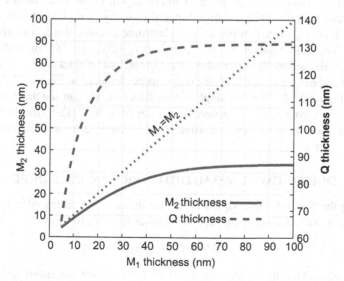

Figure 12.22 Calculated thickness for M_2 and Q as a function of the bottom metal M_1 thickness. Silver is used for both metals with an index of $0.125 - j3.35$ and a dielectric index of 1.5. The reference wavelength is 550 nm.

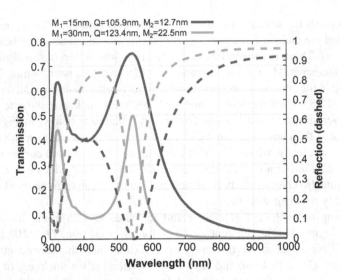

Figure 12.23 Reflection and transmission spectra of metal–dielectric–metal cavity with two different total metal thicknesses.

One of the important applications of these structures is their use in electrically conductive transparent films. In that respect, it is useful to also quantify the conductance. We will consider each metal film in the stack to be electrically shorted together. This can be done intentionally by lithographically patterning features to connect all the layers together, or most likely, it will occur unintentionally due to pin hole defects in the dielectric film. Since the quality of optical films is generally inferior to that of dielectric films used in electronic circuits, this is a reasonable assumption. The bulk conductivity of silver is $\rho = 1.59 \times 10^{-6}\Omega.\text{cm}$, and the sheet resistance is defined as $R_s = \rho/t$ where t is the total silver thickness. Using this, we can plot the sheet resistance for the design space depicted in Figure 12.22. This is shown in Figure 12.24. For this single-cavity structure, we can see that the sheet resistance ranges from $10\Omega/\square$ for very thin silver films to $0.1\Omega/\square$ for thicker films. As noted earlier, the lower sheet resistance will result in a lower peak transmission at the reference wavelength.

12.5.2 COUPLED-CAVITY METAL–DIELECTRIC BANDPASS DESIGN

Following the principles of coupled-cavity bandpass filters discussed in Section 9.3 of Chapter 9, we can couple two M_1QM_2 cavities together to form

$$M_1QM_2 \, M_1QM_2.$$

Since the effective reflectance index of M_1QM_2 made one circuitous route starting and ending at the same point, the combined structure will make two circuitous routes along the same contour. The single-cavity structure discussed in Figure 12.18 had $M_1 = 15$ nm, $Q = 105.9$ nm, and $M_2 = 12.7$ nm. The two-cavity system will be as

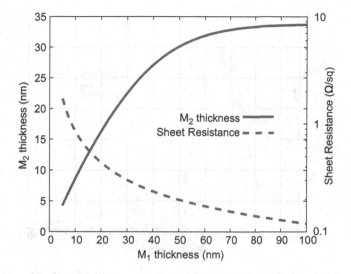

Figure 12.24 Calculated thickness for M_2 and sheet resistance as a function of the bottom metal M_1 thickness.

shown in Figure 12.25 where the adjacent metal layers have been combined into a single layer. Because we expect the effective reflectance index to be identical to the single-cavity, the reflection will still be zero at the reference wavelength. However, we can expect the transmission to be lower because there will be more metal in the two-cavity structure.

Figure 12.26 shows the transmission spectra of one-, two- and three-cavity coupled systems. The peak transmission declines from 75% in the singe-cavity structure to 69.2% and 57.7% in the two- and three-cavity coupled systems. We can also see that the transition slopes increase as the number of cavities are increased, which is consistent with the behavior of dielectric line filters. But the material dispersion has a more dominant effect compared to the coupling effect. Therefore, we can conclude that the primary effect of coupling the cavities is the reduction in the overall transmission. This may seem counterproductive, but the benefit is the decrease in sheet resistance as well as the transmission outside the desired band. The calculated

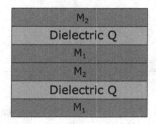

Figure 12.25 $M_1 Q M_2 M_1 Q M_2$ two-cavity coupled structure.

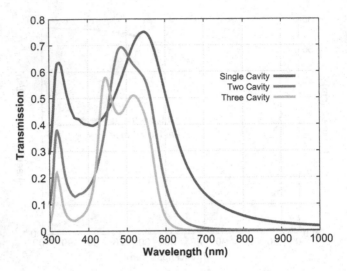

Figure 12.26 Transmission spectra of one-, two-, and three-cavity Ag–dielectric–Ag coupled systems.

sheet resistances are 0.576, 0.287, and 0.143$\Omega/\square$ for the one-, two-, and three-cavity structures, respectively. Furthermore, due to the increase in the imaginary part of the refractive index with wavelength, the cavities with more metal films exhibit a stronger reflection at longer wavelengths. They basically function as a low-cost short-pass filter or hot mirror. As a result, these structures are widely used on energy-efficient windows. Known as "low-E" coatings due to their low infrared emissivity, they can allow visible radiation to be transmitted but strongly reflect infrared radiation. This can help to reduce the thermal load in architectural windows.

Figure 12.27a shows the peak value of the transmission (which occurs at the resonance wavelength of 550 nm) as a function of the metal thickness. As expected,

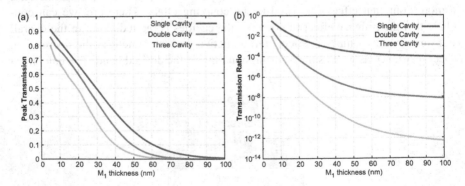

Figure 12.27 Characteristics of the one-, two-, and three-cavity structures with silver and SiO$_2$. (a) Peak transmission near 550 nm. (b) Ratio between the peak transmission and the transmission at $\lambda = 1,000$ nm.

the peak transmission declines with increasing metal thickness. However, it is interesting to note that the peak transmission does not linearly scale with the number of cavities, despite the linear increase in metal content. For example, using $M_1 = 15$ nm, the single cavity has a peak transmission of 75%, and the double cavity has a peak transmission of 69.3%. Therefore, while the sheet resistance decreases by a factor of 2, the peak transmission has only decreased to 92% of the single-cavity value. Therefore, this approach provides an attractive means of increasing optical transparency in conductive structures. Figure 12.27b shows the ratio of the peak transmission (near 550 nm) and the transmission at 1,100 nm. As expected, the multicavity configuration has a much larger ratio and can function as a more effective infrared blocking filter.

12.6 PROBLEMS

1. Design an antireflection coating for a InSb substrate at a wavelength of 1,064 nm. The refractive index of InSb at this wavelength is $4.13 - j0.19$.
2. Consider a gold–silicon–gold sandwich structure on a silica glass substrate. The top gold thickness is fixed at 20 nm. Find the bottom gold thickness and the silicon thickness to ensure resonant transmission at a wavelength of 600 nm. Then plot the reflection, transmission, and absorption spectra.
3. Design a silver–Nb_2O_5–chromium coupled-cavity structure such that the transmission band is centered at 550 nm and the ratio between the transmission at 550 nm and transmission at 750 nm is at least 100.

FURTHER READING

Dobrowolski, J. A., Li, L. & Kemp, R. A. Metal/dielectric transmission interference filters with low reflectance. 1. Design. *Applied Optics* **34**, 5673–5683 (September 1995).

Kats, M. A., Blanchard, R., Genevet, P. & Capasso, F. Nanometre optical coatings based on strong interference effects in highly absorbing media. *Nature Materials* **12**, 20–24. ISSN: 1476-4660 (2013).

Lemarquis, F. & Marchand, G. Analytical achromatic design of metal-dielectric absorbers. *Applied Optics* **38**, 4876–4884 (Auguest 1999).

MacLeod, H. A. A new approach to the design of metal-dielectric thin-film optical coatings. *Optica Acta: International Journal of Optics* **25**, 93–106 (1978).

Mu, J. et al. Design and fabrication of a high transmissivity metal-dielectric ultraviolet band-pass filter. *Applied Physics Letters* **102**, 213105 (May 2013).

Sarangan, A. Design of metal-dielectric resonant-cavity thin-film structures using the effective reflectance index method. *The Journal of the Optical Society of America B* **35**, 2294–2301 (September 2018).

Sullivan, B. T. & Byrt, K. L. Metal/dielectric transmission interference filters with low reflectance. 2. Experimental results. *Applied Optics* **34**, 5684–5694 (September 1995).

13 Thin-Film Designs Using Phase Change Materials

13.1 INTRODUCTION

In the general literature, phase change materials (PCMs) are those that change between solid, liquid, and vapor phases and produce large changes in energy during the transition. However, in photonics and in electronics, PCMs are understood as materials that remain in the solid phase but change some other characteristic, such as crystal structure or valence state, and produce large changes in their optical constants. The ability to tune the refractive index of a material by temperature, voltage, or some other means opens up many applications that cannot be realized using regular materials whose refractive index is fixed or changes by only a small amount.

Every material will exhibit a change in refractive index due to an external stimulus, such as temperature, pressure, or electric fields. These are quantified through their thermo-optic, piezo-optic, and electro-optic coefficients, respectively. A small coefficient is a good thing when we are seeking environmental stability, but a large coefficient is desirable when designing a sensor or tunable device. Pressure and temperature sensors, for example, exploit the thermo-optic and piezo-optic effects. Devices such as Mach–Zehnder modulators use the electro-optic effect to induce an optical phase change, which is then combined with another wave to induce a change in interference pattern at the output port. Additionally, laser cavities and line filters are sometimes tuned by controlling their temperature. However, the change in refractive index in all of these examples is very small. For example, the thermo-optic coefficient dn/dT of fused silica is $\sim 10^{-5}/°C$ in the visible spectrum. This means a $100°C$ change in temperature will produce a refractive index change of only 0.001. Similarly, silicon has a thermo-optic coefficient of 10^{-4} at 1,550 nm. These effects are so small that they become observable only in highly resonant cavities, such as the resonance-enhancement cavities discussed in Chapter 9. Their effects in typical optical thin-film devices are usually too small to be of practical use.

Unlike the aforementioned optical effects, PCMs produce more dramatic changes in their electrical, optical, and mechanical properties. They undergo a physical change, such as a change in crystal or molecular structure. This results in a change in refractive index that is many orders of magnitude larger than that in conventional materials.

For example, vanadium dioxide (VO_2) will switch from a monoclinic to a tetragonal crystal structure when raised above $68°C$ temperature. Other PCMs can be crystallized or randomized (amorphisized) by the application of heat. $Ge_2Sb_2Te_5$ is a chalcogenide glass that can be crystallized by raising its temperature above $150°C$.

In this chapter, we will discuss the principles of thin-film designs involving PCMs. The design principles are largely the same as the ones that were discussed in earlier chapters since most PCMs can be treated as dielectrics, metals, or some combination thereof. However, they still warrant a separate discussion because of the unique functionalities that can be achieved.

13.2 VANADIUM DIOXIDE (VO$_2$)

13.2.1 OPTICAL PROPERTIES OF VANADIUM DIOXIDE

Vanadium is a transition metal with many different oxidization states. Several of these oxides exhibit some type of phase change. Among them, vanadium dioxide (VO$_2$) has garnered the most attention because its transition temperature is close to room temperature. When heated above 68°C, VO$_2$ undergoes an abrupt change in its crystal structure. The room temperature phase has a monoclinic crystal structure and behaves as a semiconductor. Above 68°C, it switches to a tetragonal crystal structure and exhibits characteristics of a metal. When cooled down, it reverts back to its original low-temperature phase. Hence, this is a reversible transition. During this transition, the values of n and κ as well as the electrical conductivity undergo large changes. These are shown in Figures 13.1 and 13.2.

As can be seen, the largest change in refractive index occurs in the infrared spectrum above 1 μm. In the cold state (25°C), the refractive index is between 2.0 and 2.5 with an imaginary part that declines with increasing wavelength. In the hot state (80°C), both the real and imaginary parts increase with wavelength. The real part of its permittivity becomes negative at a wavelength of around 1.0 μm (where $m = \kappa$) and continues to become more negative as wavelength increases. As discussed in

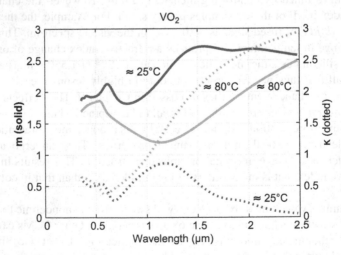

Figure 13.1 Refractive index of VO$_2$ above and below the transition temperature. The dark lines are for the cold state, and the gray lines are for the hot state (Guo et al., in press).

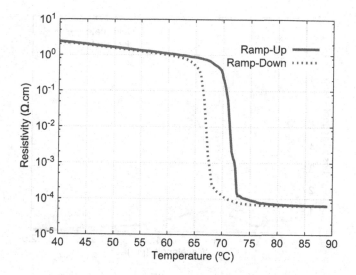

Figure 13.2 Resistivity of VO_2 as a function of temperature.

Chapter 12, a negative value for permittivity is characteristic of metals; hence this transition is also known as insulator-to-metal transition (IMT).

It should be noted that the properties of VO_2 vary quite a bit depending on growth conditions. The results shown in Figures 13.1 and 13.2 were taken from VO_2 films grown in the author's own laboratory.

The behavior of resistivity versus temperature is shown in Figure 13.2. We can see a sharp decrease in resistivity at around 70°C by nearly four orders of magnitude, which arises due to the IMT. During cool-down, the resistivity remains low until the temperature falls below 65°C, at which point the resistivity increases back to its original high value. This hysteresis is a well-known behavior in VO_2 that lends itself to be utilized as a memory element.

The loss tangent of VO_2 as a function of wavelength is shown in Figure 13.3. We can see that the cold state has low loss tangent values. As a result, it behaves reasonably similar to a dielectric, especially at longer wavelengths. In the hot state, the loss tangent becomes large, spiking to a very large value between 1.0 and 1.5 μm. In some sense, the hot state of VO_2 resembles the loss tangent of chromium (see Figure 12.2 in Chapter 12). Therefore, near 1.0 μm, VO_2 can be used as a variable absorber by switching between the cold and hot states, and at wavelengths longer than 2.0 μm, it can be used as a variable reflector. We will explore both of these aspects in this chapter.

13.2.2 ANTIREFLECTION

For our first example, let's consider the simplest case of creating antireflection for a VO_2 film on a sapphire substrate. VO_2 is typically grown on sapphire because the lattice constants of sapphire (Al_2O_3) and VO_2 are closely matched; hence the

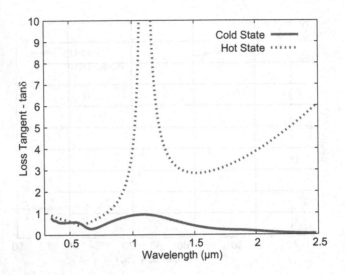

Figure 13.3 Loss tangents of VO_2 in the hot (80°C) and cold (25°C) states.

highest-quality VO_2 is obtained on these substrates. Let's consider a reference wavelength of 2.0 μm. The refractive index of sapphire at 2.0 μm is 1.73, and the refractive index of VO_2 in the cold state is $2.67 - j0.23$. The thickness of VO_2 can, of course, be any value, but for this example, let's consider it to be 500 nm. Using the techniques discussed in Section 5.6 of Chapter 5, we can calculate the required film to produce antireflection at a wavelength of 2.0 μm. The solution is a film index of 1.76 with a thickness of 295 nm. The contour plot for this case is shown in Figure 13.4. When the VO_2 film is in the cold state, the contour of the top film ends at $n_r = 1.0$. When the VO_2 switches to the hot state, the refractive index of VO_2 becomes $2.04 - j2.64$. As discussed in Chapter 12, the large imaginary part drives the contour toward its refractive index value of $2.04 - j2.64$. The top film then brings the final effective reflectance index to $n_r = 0.60 + j0.86$. This results in a reflection of 27%.

Just as in the case with metal films, the reflection does not tell us anything about the transmission. This has to be computed using the transfer matrix method (TMM). Figure 13.5 shows the reflection and transmission from this structure when the VO_2 is in the cold state as well as in the hot state. As expected, the reflection falls to zero at a wavelength of 2,000 nm when VO_2 is in the cold state. The transmission is around 47%. In the hot state, the reflection rises to 27%, and the transmission drops to nearly zero. In other words, the absorption in the VO_2 film is 53% in the cold state and 73% in the hot state. The most common application of this type of film stack is in the area of optical limiting. Optical limiting is the phenomenon where the light transmission through a device declines with increasing light intensity. The intrinsic absorption of VO_2 in the cold state (53% in the above example) will lead to a temperature rise. The more intense the beam, the higher the temperature. When the temperature rises above 68°C, the VO_2 will switch to the hot state, and the transmission will decline

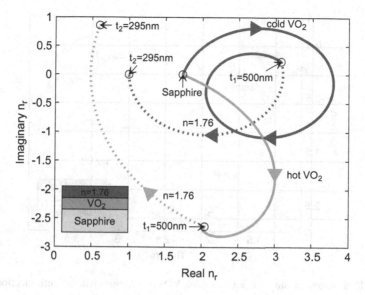

Figure 13.4 Contour plot of a 500 nm thick VO_2 on sapphire with an antireflection coating film ($n_f = 1.76$ and $t_2 = 295$ nm). The antireflection coating is designed such that the system produces zero reflection in the VO_2 cold state.

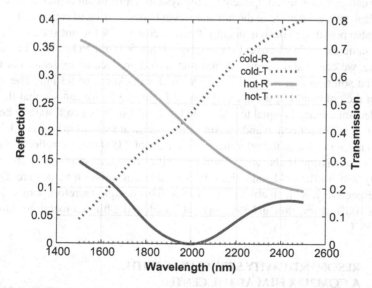

Figure 13.5 Reflection and transmission spectra from a 500 nm thick VO_2 on sapphire with an antireflection coating film ($n_f = 1.76$ and $t_2 = 295$ nm). The dark lines are for the cold state, and the light lines are for the hot state. The film stack produces antireflection in the cold state.

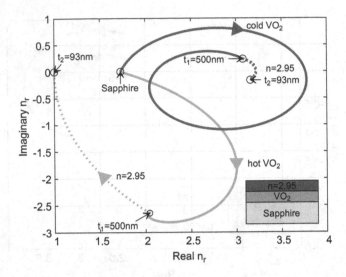

Figure 13.6 Contour plot of a 500 nm thick VO_2 on sapphire with an antireflection coating film ($n_f = 2.95$ and $t_2 = 93$ nm). The antireflection coating is designed such that the system produces zero reflection in the VO_2 hot state.

to nearly zero. Hence, this feature can be used to block high-energy beams from causing damage to other components in the system. Unfortunately, however, the VO_2 film will absorb even more in the hot state, eventually leading to catastrophic failure.

It is also possible to design an antireflection coating for the hot state of VO_2. Using the same example discussed previously, with a 500 nm VO_2 film on a sapphire substrate, we can find the required film that would produce antireflection in the hot state. The solution is a film index of 2.95 and a thickness of 93 nm. The contour plots for the cold and hot states are shown in Figure 13.6. We can see that the effective reflectance index is equal to 1.0 in the hot state, but in the cold state, it becomes $3.15 - j0.17$. The reflection and transmission spectra are shown in Figure 13.7. What is worth noting is that at the reference wavelength of 2,000 nm, the reflection is zero, and the transmission is also zero. Therefore, all of the incident power is absorbed in the VO_2 film. In the cold state, the reflection and transmission values are 27% and 34%, respectively, with an absorption of 39%. This design, therefore, can switch between a 100% absorption and 39% absorption when the film temperature transitions across 68°C.

13.2.3 RESONANT-CAVITY STRUCTURES WITH A COMPLEX FILM AT THE CENTER

In applications where the switching is initiated by the incident light beam via light absorption (resulting in a temperature rise), it is useful to have a means to control the light intensity when this switching will take place. Part of this lays in the thermal

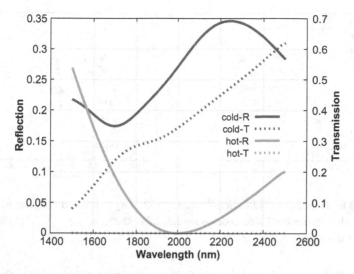

Figure 13.7 Reflection and transmission spectra from a 500 nm thick VO$_2$ on sapphire with an antireflection coating film ($n_f = 2.95$ and $t_2 = 93$ nm). The dark lines are for the cold state, and the light lines are for the hot state. The film stack was designed to produce antireflection in the hot state.

conductivity from the VO$_2$ film to the surrounding medium. We will not discuss that aspect here. The second aspect is the field intensity in the VO$_2$ film. We can enhance the field intensity in the VO$_2$ film by placing it at the center of a resonant cavity. These structures (line filters) were examined in Chapter 9. We showed that absorption in the central quarter-wave film can be increased manyfold due to the resonance effect. In Chapter 9, we introduced an imaginary part for the central quarter-wave film. We also argued that the resonance of the cavity will not be significantly altered as long as the imaginary part is very small. Clearly, the imaginary part of VO$_2$ is not insignificant, whether in the cold state or in the hot state. In this section, we will do a more systematic development of a resonant cavity with a complex material at the center that is applicable to PCMs.

Figure 13.8a shows the contour plot of the $\frac{H}{2} R_{3H} H R_{3H} \frac{H}{2}$ line-cavity structure. This is the same as Figure 9.2a from Chapter 9, but this figure also includes additional contours for complex-valued center layers. All of the lines start at the effective reflectance index value of $n_r = 1.0$. The first arc is due to the first $\frac{H}{2}$ layer. This is followed by three unit cells of $\left(\frac{H}{2} L \frac{H}{2} \right)$ that bring the effective index asymptotically toward the final value of jn_1 (which in this case is $j2.5$). The effective reflectance index value at the end of the three unit cells is $0.093 + j2.49$. The central quarter-wave H layer (for $n = 2.5$) is a large arc that covers almost the full circle because it spans the lower-density portion of the arc (see Figure 5.7 in Chapter 5 for explanation). It brings the index value to a symmetric location on the lower side of the plot. It's value is $0.093 - j2.49$. In other words, the starting and ending values of the central H when $n = 2.5$ are complex conjugates of each other. The final three unit cells of

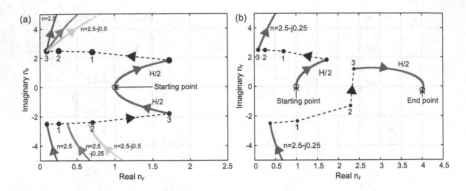

Figure 13.8 Contours of $\frac{H}{2}R_{3H}IR_{3H}\frac{H}{2}$ with $n_1 = 2.5$, $n_2 = 1.5$, and substrate $n_r = 1.0$ (a) Contours when the central I layer indexes are $2.5, 2.5 - j0.25$, and $2.5 - j0.5$. (b) Full contour when the I layer index is $2.5 - j0.25$.

$\left(\frac{H}{2}L\frac{H}{2}\right)$ and the last $\frac{H}{2}$ layer bring the final effective reflectance index to the starting point of $n_r = 1.0$.

Now, if the refractive index of the central H layer has a small imaginary part (as a result, we will label it I instead of H), its ending value will inevitably not be at the same location. In this case, we are selecting the I layer thickness such that the real part of its phase is one quarter wave $(\frac{\pi}{2})$. We know from our discussion on metals in Chapter 12 that a contour of a complex index film will trace a circle that shrinks in diameter and converges toward its material index value. Therefore, we can predict that the ending value of the central I layer will be a little to the right of the original value. This is exactly what we observe in Figure 13.8a. When the refractive index of the I layer is $2.5 - j0.25$, we can see that the ending point of the arc moves a little to the right. When the refractive index is $2.5 - j0.5$, the point moves even further to the right. To avoid clutter, we have not plotted the preceding three unit cells for these contours. Clearly, the final end point of these contours will not be at $n_r = 1.0$. Hence, we cannot expect them to be resonant at the reference wavelength. Figure 13.8b shows the full contour of $\frac{H}{2}R_{3H}IR_{3H}\frac{H}{2}$ when the I layer index is $2.5 - j0.25$, where we can see that the end point is significantly far from the desired end point of $n_r = 1.0$. However, we should be able to deduce from Figure 13.8b that we may be able to recover the end point to some extent if we select a fewer number of unit cells in the trailing edge of the structure such as, for example, $\frac{H}{2}R_{3H}IR_{2H}\frac{H}{2}$. This is exactly the approach we will take to redesign the cavity containing an absorptive layer. Absorption makes the cavity asymmetric; i.e., even with symmetric reflection, the reflected field will be smaller than the incident field. We can restore the symmetry by increasing the reflectivity of the front mirror compared to the rear mirror. In other words, a higher reflection of the front mirror when combined with the absorption can become equal to the rear mirror. Unfortunately, however, the mirror reflectivities cannot be tuned to arbitrary reflectivities and phases. Since the number of unit cells has to be an integer, we are limited only to certain discrete values. This mismatch can be corrected by splitting the I layer into two films.

Consider, for example, $\frac{H}{2}R_{3H}IR_{2H}\frac{H}{2}$. From the above discussion, we know that it will be closer to resonance than the symmetric structure $\frac{H}{2}R_{3H}IR_{3H}\frac{H}{2}$. Using $n_1 = 2.5$ and $n_2 = 1.5$, we can calculate the end point after the three unit cells to be $0.093 + j2.498$. The desired end point after the I layer can also be determined by calculating the complex conjugate of the end point of $\frac{H}{2}R_{2H}$. This works out to be $0.258 - j2.486$. Now we need to find the thicknesses of a two-film structure consisting of $n_{f1} = 2.5 - j0.25$ and $n_{f2} = 2.5$ such that these two points become connected. That is,

$$n_{r3} = n_{f1} \frac{(n_{r1} + n_{f1}) + (n_{r1} - n_{f1})e^{-j2\theta_1\left(1 + j\frac{\kappa_{f1}}{m_{f1}}\right)}}{(n_{r1} + n_{f1}) - (n_{r1} - n_{f1})e^{-j2\theta_1\left(1 + j\frac{\kappa_{f1}}{m_{f1}}\right)}}, \tag{13.1}$$

where n_{r1} is the starting point (end of the three unit cells), n_{f1} is the complex film, and θ_1 is the real part of the phase thickness. The second film starts from n_{r1} and ends at n_{r2} such that

$$n_{r2} = n_{f2} \frac{(n_{r1} + n_{f2}) + (n_{r1} - n_{f2})e^{-j2\theta_2}}{(n_{r1} + n_{f2}) - (n_{r1} - n_{f2})e^{-j2\theta_2}}, \tag{13.2}$$

where n_{r2} is the dielectric film and θ_2 is its phase thickness. The condition that needs to be satisfied is n_{r2} be equal to the known end point (complex conjugate of trailing edge of the cavity). In this case, $n_{r1} = 0.093 + j2.498$, $n_{f1} = 2.5 - j0.25$, $n_{f2} = 2.5$, and $n_{r2} = 0.258 - j2.486$. As discussed in Chapter 12, this condition really consists of two equations for the real and imaginary parts separately, which allows us to uniquely solve for θ_1 and θ_2.

The solution is shown in Figure 13.9. The calculated values are $\theta_1 = 0.77\frac{\pi}{2}$ and $\theta_2 = 0.23\frac{\pi}{2}$, which correspond to film thicknesses of 42.4 and 12.6 nm, respectively, using a reference wavelength of 550 nm. We can see that the points between the leading and trailing unit cells are correctly linked by the I layer (which really consists of two films). As a result, we can expect the reflection to fall to zero at the reference wavelength.

Figure 13.10 shows the reflection and transmission spectrum from this structure calculated using the TMM. Clearly, the resonance can be seen at the reference wavelength. We can also see that the peak transmission at the resonance is about 35%, which means the absorption at this wavelength is 65%. The single pass absorption without any interfacial reflections through a 42.4 nm thick film of index $2.5 - j0.25$ is 21.5% at a wavelength of 550 nm. Therefore, this resonant cavity enhances the absorption by roughly three times. This enhancement factor can be increased by using a larger number of unit cells. For example, using five unit cells on the incident side and four unit cells on the transmission side would result in just 1.57 and 53.4 nm for the $2.5 - j0.25$ and 2.5 index layers. The peak transmission is 36%, which is a 40-fold increase in absorption due to the resonance-cavity enhancement (RCE). Needless to say, various combinations of unit cells and I-layer indices can be utilized to produce this resonance-cavity enhancement.

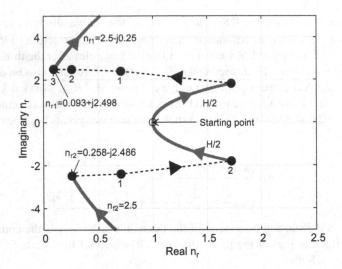

Figure 13.9   Contour of $\frac{H}{2}R_{3H}IR_{2H}\frac{H}{2}$ with $n_1 = 2.5$, $n_2 = 1.5$, and substrate $n_r = 1.0$, solved using two films ($n_{f1} = 2.5 - j0.25$ and $n_{f2} = 2.5$) in the I layer.

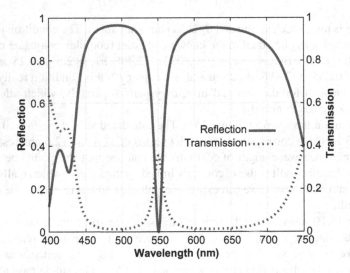

Figure 13.10   Spectral reflection and transmission for the $\frac{H}{2}R_{3H}IR_{2H}\frac{H}{2}$ structure shown in Figure 13.9.

13.2.4   RESONANT-CAVITY STRUCTURES WITH VO₂ AT THE CENTER

Returning to the VO$_2$ thin film, we can place it in the center of a cavity as $\frac{H}{2}R_{5H}IR_{4H}\frac{H}{2}$ to increase resonant absorption and to reduce the light intensity required to induce switching. Let's consider the H and L layers to be 2.5 and 1.5 at a reference wavelength of 2,000 nm. The VO$_2$ cold and hot indices are $2.04 - j2.64$

and $2.67 - j0.23$, respectively, at this wavelength. We will consider the central I layer to consist of a VO_2 thin film and another dielectric with an index of 2.5. Using the same solution technique described earlier, the thicknesses of the VO_2 and the dielectric can be solved to be 5.9 and 193.7 nm, respectively. The calculated reflection and transmission spectra for the $\frac{H}{2}R_{3H}IR_{2H}\frac{H}{2}$ structure are shown in Figure 13.11 for both the cold and hot VO_2 phases. The cavity structure is in resonance when the VO_2 film is in the cold state. As a result, the reflection falls to zero at the reference wavelength. The transmission is about 35%, with the remainder (65%) absorbed due to resonance-enhanced absorption. For comparison, the absorption in a free-standing 5.9 nm VO_2 will be 0.85%. Therefore, the resonant effect has increased the absorption by 75-fold. This would effectively reduce the light intensity required to induce switching by the same amount, assuming, of course, that the thermal conductivity of the structure remains the same. When the VO_2 switches to the hot state, the transmission declines to 2.8%, and the reflection increases to 52%. The transmission switching ratio is about 20. Even though the structure is no longer in perfect resonance, the peak transmission wavelength moves to a slightly shorter wavelength of 1,990 nm.

Clearly, many combinations of reflector unit cells are possible, resulting in different RCE and switching ratios. Some of these are listed in Table 13.1. We can make the observation that the RCE of absorption increases as the cavity strength increases.

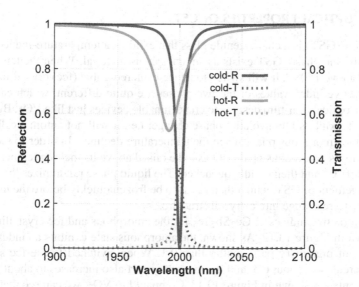

Figure 13.11 Spectral reflection and transmission for the $\frac{H}{2}R_{3H}IR_{2H}\frac{H}{2}$ structure using a VO_2 film and a 2.5 index dielectric film with thicknesses of 5.9 and 193.7 nm for the VO_2 film and dielectric of the I layer, respectively.

Table 13.1

Resonance Conditions Calculated for Different Unit Cell Combinations of $\frac{H}{2}R_{NH}IR_{MH}\frac{H}{2}$ Using 2.5 and 1.5 as the High and Low Film Indices and the I Layer Consisting of a VO_2 Film and a Dielectric Layer with Index of 2.5

Unit Cells (M/N)	VO_2 Thickness (nm)	Dielectric Thickness (nm)	Peak Transmission (Cold)(%)	RCE	Peak Transmission (Hot) (%)	Extinction Ratio
2/1	230.43	363.15	35.8	1.26	2.0	17.7
3/2	160.16	36.76	34.2	1.65	2.6	12.8
3/1	254.07	336.93	13.3	0.43	0.4	33.1
4/3	18.7	180.2	35.3	13.2	2.7	12.7
4/2	182.3	14.0	12.6	0.54	0.8	14.3
4/1	262.1	327.8	4.8	0.15	0.12	39.2
5/4	5.9	193.7	35.7	42.1	2.9	12.3
5/3	27.7	170.8	12.6	3.2	0.62	20.2
5/2	188.3	7.7	4.5	0.3	0.19	15
5/1	265.0	324.5	1.7	0.05	0.04	40.1
6/5	2.04	197.8	35.8	121.5	2.9	12.2

The resonance condition is calculated for the cold state of VO_2.

13.3   GE$_2$SB$_2$TE$_5$ (GST)

13.3.1   OPTICAL PROPERTIES OF GST

Ge$_2$Sb$_2$Te$_5$ (GST) is a chalcogenide glass that exhibits a temperature-induced phase change. In one phase, GST exists as an amorphous material. When the temperature is raised above $150°C$, it will switch to a face-centered-cubic (fcc) crystalline phase. The refractive index values of the two phases are quite different, which can be exploited in optical thin-film designs to create tunable devices just like VO_2. But unlike VO_2, GST exhibits a nonvolatile phase change; i.e., it will not automatically revert back to the amorphous phase when the temperature declines. In order to switch the material to its amorphous phase, it has to be raised above its melting temperature (at around $650°C$) and then rapidly quenched. The liquid phase randomizes the molecular arrangement of GST, which then needs to be frozen quickly before the molecules have time to rearrange into a crystal structure.

The refractive indices of Ge$_2$Sb$_2$Te$_5$ for the amorphous and fcc crystalline states are shown in Figure 13.12. As shown, the amorphous state exhibits an index around 4.5 with an imaginary part smaller than 0.1. When switched to the fcc state, the index increases to about 6.5, and the imaginary part also increases to about 1.0. The loss tangents are shown in Figure 13.13. Compared to VO_2, we can see that the loss tangents are much lower in GST in both states. The real part of the refractive index shifts by almost 2.0 during the phase transition. This makes GST a better choice for designing low loss tunable devices. It should also be mentioned that there are many

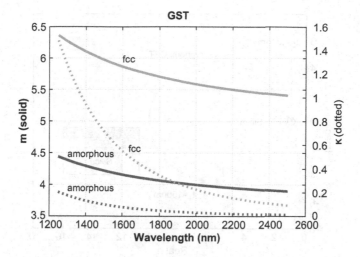

Figure 13.12 Refractive index of GST in the amorphous and crystalline states.

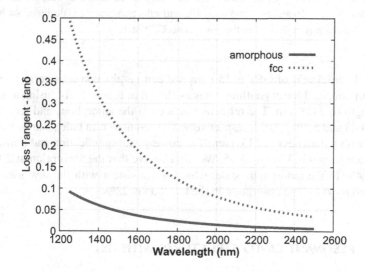

Figure 13.13 Loss tangents of GST in the amorphous and crystalline states.

other stoichiometries of Ge, Sb, Te which exhibit different properties. Most notably, GSST ($Ge_2Sb_2Se_4Te_1$) has been shown to have lower optical losses than GST.

13.3.2 ANTIREFLECTION

Antireflection designs using GST are quite straightforward, and the procedure is identical to the methods discussed for VO_2 in Section 13.2.2. For completeness, we will briefly consider the example of a GST film on a silica glass substrate. If we

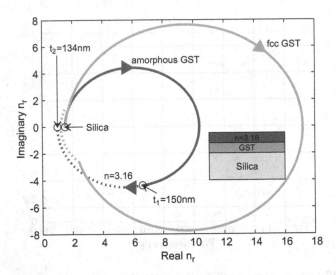

Figure 13.14 Contour of a 150 nm thick GST on a silica substrate with an antireflection coating film ($n_f = 3.16$ and $t_2 = 295$ nm). The antireflection coating is designed such that the system produces zero reflection in the amorphous GST state.

assume the thickness of GST is 150 nm, we can get the contours corresponding to the amorphous and fcc crystalline states as shown in Figure 13.14, using a reference wavelength of 2,000 nm. The refractive indices of the amorphous and fcc states are $4.0 - j0.03$ and $5.59 - j0.22$, respectively. The required film index can be calculated as 3.16 with a thickness of 133.6 nm. The corresponding reflection and transmission spectra are shown in Figure 13.15. We can observe that the absorption at $2,000$ nm is only 4% in the amorphous state, which is consistent with the low loss tangent compared to VO_2. The absorption in the fcc state is 22%.

13.3.3 RESONANT-CAVITY STRUCTURES WITH GST

Just like with VO_2, GST can be placed inside a cavity to produce a resonant transmission. If the cavity is designed to be in resonance for the amorphous state, when the GST switches to the fcc state, the resonance will be dramatically altered. Figure 13.16 shows an example of a $\frac{H}{2}R_{3H}IR_{2H}\frac{H}{2}$ where the high and low indices are 2.5 and 1.5, and the I layer consists of a GST layer and 2.5-index dielectric. Using a reference wavelength of 2,000 nm, we can solve for the GST thickness to be 234.9 nm and the dielectric thickness to be 230.6 nm. The peak transmission is 43% in the amorphous state at the reference wavelength and declines to 0.2% in the fcc state. In fact, we can notice that the resonance has moved to a longer wavelength near $2,250$ nm due to the higher refractive index in the fcc state. The resonance has also become weaker due to the higher absorption in the fcc state.

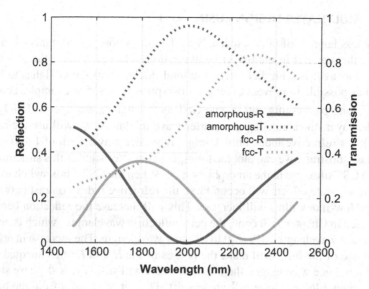

Figure 13.15 Reflection and transmission spectra from a 150 nm thick GST on a silica substrate with an antireflection coating film ($n_f = 3.16$ and $t_2 = 133.6$ nm). The solid lines are for the amorphous state, and the dashed lines are for the fcc state. The film stack was designed to produce antireflection in the amorphous state.

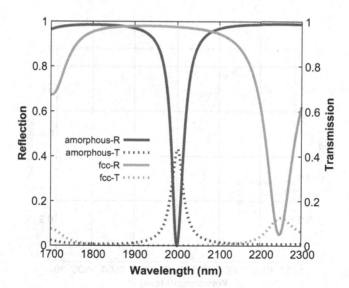

Figure 13.16 Spectral reflection and transmission for the $\frac{H}{2} R_{5H} I R_{4H} \frac{H}{2}$ structure using a GST film and a 2.5 index dielectric film with thicknesses of 234.9 and 230.6 nm for the VO_2 film and dielectric of the I layer, respectively.

13.3.4   MULTILAYER DESIGNS USING GST

The low loss tangent of GST, especially in the amorphous phase, makes it possible to utilize the material in multilayer configurations. However, since the loss tangent is not as low as it is with dielectrics, the total thickness of the GST has to be kept as small as possible to prevent excessive absorption losses. For example, consider a $\left(\frac{H}{2}L\frac{H}{2}\right)^N$ long-pass configuration with GST as the high-index layer and SiO_2 as the low-index layer. In calculating the quarter-wave thickness, we will use the real part of the GST's refractive index only. Using a reference wavelength of 1,400 nm on a silica glass substrate, we can plot the long-pass characteristics of this structure when all of the GST films are in the amorphous state. When the GST films switch to the fcc state, two separate effects will occur. First, the refractive index contrast between the high- and low-index films will increase. This will increase the reflection bandwidth as discussed in Chapter 7. Second, the peak reflection wavelength (which is normally the reference wavelength) will shift to a longer wavelength. The increase in refractive index of one of the films will make the phases of the H and L films unequal. At the original reference wavelength, the low-index film will still have a $\frac{\pi}{2}$ phase shift, but the high-index film will have $>\frac{\pi}{2}$ phase shift. The net phase of π from the high and low index films will now occur at a smaller k_0 (or at a longer wavelength). Hence the peak reflection wavelength will shift to a longer wavelength. The details of this analysis is left as an exercise for the reader. The end result is that a red shift of the long-pass edge is observed as the refractive index of the H film increases in value. This is shown in Figure 13.17.

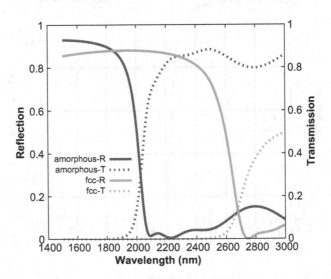

Figure 13.17   Spectral reflection and transmission of $\left(\frac{H}{2}L\frac{H}{2}\right)^6$ using GST as the high-index film and 1.5 for the low-index film.

FURTHER READING

Guo, P., Sarangan, A. & Agha, I. A review of Germanium-Antimony-Telluride phase change materials for non-volatile memories and optical modulators. *Applied Sciences* **9**, 530. ISSN: 2076-3417 (Feburary 2019).

Guo, P. et al. Vanadium dioxide phase change thin films produced by thermal oxidation of metallic vanadium. *Thin Solid Films* (in press). doi: 10.1016/j.tsf.2020.138117

Kocer, H. et al. Intensity tunable infrared broadband absorbers based on VO2 phase transition using planar layered thin films. *Scientific Reports* **5**, 13384. ISSN: 2045- 2322 (2015).

Sarangan, A. et al. Broadband reflective optical limiter using GST phase change material. *IEEE Photonics Journal* **10**, 1–9. ISSN: 1943-0655 (April 2018).

Taha, M. et al. Insulator-metal transition in substrate-independent VO(2) thin film for phase-change devices. *Scientific Reports* **7**. PMC5738395[pmcid], 17899–17899. ISSN: 2045-2322 (December 2017).

14 Deposition Methods

14.1 INTRODUCTION

14.1.1 OPTICAL THIN-FILM DESIGN VS PROCESS DESIGN

Even though we are discussing the deposition aspects of optical thin-film structures at the very end of this book, in practice, deposition methods and their limitations must be considered before the design process, not after the fact. Before a design can even begin, the film materials and appropriate deposition methods must be chosen. This selection is based on wavelength range, expected film thickness range, substrate temperatures, material stability as well as compatibility with other materials in the structure.

Unless there are compelling reasons otherwise, deposition should be done using a single technique in a single deposition chamber by alternating the layer materials. Switching between equipment, especially if they require breaking vacuum, would significantly add to the complexity and cost of a project.

14.1.2 MAJOR CATEGORIES OF DEPOSITION TECHNIQUES

Thin-film deposition techniques can be broadly categorized into physical vapor deposition (PVD) and chemical vapor deposition (CVD). There is a third method that uses liquid spin coating, which does not involve vaporization, but it is not widely used for making optical films, so we will not discuss it here. In both PVD and CVD methods, the material or precursors to be deposited arrive in vapor form and condense on the substrate (hence the term "vapor deposition"). In PVD, the material to be deposited is vaporized by a physical process, such as the application of heat, ion bombardment, or a laser pulse. In CVD, gaseous precursors are used to produce a chemical reaction that results in the desired thin film. Both methods are performed in a vacuum chamber to minimize the effects of atmospheric contaminants. The majority of optical thin films are made using physical vapor deposition methods because they are simpler and can be used with a wider range of materials. CVD is used in applications where film quality, conformal coverage, and stoichiometry control are more important. Each method has benefits and disadvantages, and it is important to understand these trade-offs. A summary of these deposition methods is illustrated in Figure 14.1.

14.2 PVD

PVD employs energy sources such as ions, heat, or laser pulses to atomize a source material. Sputter deposition uses ions, thermal evaporation uses heat, and pulsed laser deposition (PLD) uses laser pulses. The material to be deposited is usually in the same chemical state as the source material. For example, when depositing a thin

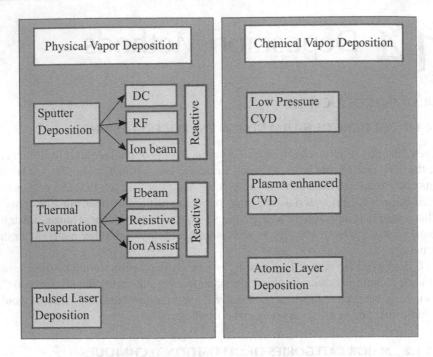

Figure 14.1 Summary of deposition methods.

film of SiO_2, the source material will also be SiO_2. However, there is a variant of PVD known as reactive PVD where chemical reaction with a gas is used to alter the composition of the source material. For example, it is possible to produce SiO_2 from a silicon source by reacting it with oxygen during the deposition.

Typical pressures used in PVD are in microtorrs (μTorr) (10^{-6} Torr) to millitorrs (mTorr) (10^{-3} Torr). In this pressure range, the deposition species behave as ballistic particles and travel in a nearly straight line from the source to the substrate. As a consequence, the deposition profile will depend on the geometry of the setup and the position of the substrate in relation to the source. Sidewalls, tilted surfaces, and off-axial areas will experience a much smaller deposition rate. To a large extent, this nonuniformity can be averaged out by rotating the substrates during deposition. However, a single-axis rotation cannot correct all these nonuniformities. Production systems generally utilize complex rotation patterns known as planetary rotations.

14.2.1 SPUTTER DEPOSITION

Also known as cathode sputtering, this method utilizes gas ions in a plasma as tiny projectiles to vaporize the source material. It is a "cold" process, in the sense that vaporization takes place via momentum transfer rather than by heating. Typically argon gas is used to produce the ions. Argon is inert and therefore will not chemically interfere with the deposition process. It is also an inexpensive gas. Argon ions are relatively heavy and have reasonably high sputter yields for commonly used deposition materials. Sputter yield is defined as the number of ejected atoms for each incident ion.

Details of the plasma configurations and vacuum systems can be found elsewhere, but the essential elements of a sputter deposition system can be summarized as the cathode, anode, and the process gas in a vacuum chamber. Anode is typically tied to the chamber walls and grounded, so that the electric potential on the cathode will be negative compared to ground. An ultrahigh vacuum system is critical for reducing contamination from background gases. Atmospheric moisture is the most important of these, because water vapor is much harder to remove than oxygen, nitrogen, and other such gases. Argon gas is fed into the chamber and ionized by the electric field between the anode and the cathode. This produces Ar^+ ions and free electrons in a fluid state known as the plasma. The ions will be attracted toward the cathode, and electrons will be attracted toward the anode. Ions being much heavier than electrons will impact the cathode with much greater momentum than the electrons. This impact will dislodge electrons and neutral atoms from the cathode. The electrons released from the cathode are necessary for sustaining the plasma. They compensate for the electrons reaching the anode, as well as those lost due to recombination in the plasma. The ejected neutral atoms are the sputtered atoms. These atoms emerge from the cathode with significant recoil energy and disperse into the surrounding area. The substrate is placed in the path of the ejected atoms such that they collide and condense on the substrate to create the desired thin film. This is the basic description of the sputter deposition process and is diagrammatically illustrated in Figure 14.2.

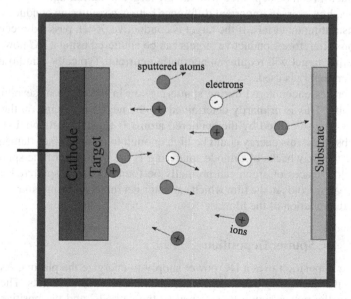

Figure 14.2 Sputter deposition process.

The source material for the deposition, known as the target material, is mechanically clamped to the cathode electrode. This allows the ions to strike the target material instead of the cathode. As a result, the target has to be in the same shape and size as the cathode. Cathode geometries come in virtually every shape imaginable, but the most common geometries used in research laboratories are 3- or 4-inch diameter circular shapes.

The background gas pressure during deposition is in the range of a few millitorrs (mTorr). This is primarily dictated by the pressure required to sustain the plasma. In addition, most cathodes utilize magnets to focus and draw the plasma toward the cathode. Known as magnetron cathodes, their basic purpose is to increase the plasma density near the cathode and increase the sputter efficiency (number of ejected atoms per unit discharge power).

Sputtering action is an atomic-level interaction between the ions and the target atoms. The target atoms are stripped from their bound states and are sputtered one atom at a time. Furthermore, the sputter yield of each element is different. Given that, it may be surprising that compounds and alloys can be sputtered just like single elements and without significant change in their stoichiometry. This has to do with the evolution of the target's surface stoichiometry. After some time, known as the target conditioning time, the surface composition of the target will evolve toward a slightly different composition than its starting stoichiometry. This composition, combined with the different sputter yields of the constituent elements, will result in the stoichiometrically correct removal of the target atoms. As a result, mixtures of materials such as NiCr and indium tin oxide (ITO), as well as compounds such as SiO_2 are commonly deposited using sputtering. Of course, elemental metals and semiconductors can be deposited without any target conditioning.

There is, however, an important difference between sputtering conductive materials and insulating materials. If the target is conductive, it can pass a direct current to the plasma. Therefore, conductive targets can be sputtered using a DC power supply. An insulating target will require an alternating current. Typically, a radio frequency (RF) power supply is used.

Ion energies encountered during sputtering are in the range of several hundred electronvolts. This is primarily determined by the negative voltage on the cathode. The recoil energy carried by the sputtered atoms is typically around 1–10 eV. On fragile substrates, this energy could be high enough to cause surface damage.

One of the drawbacks of cathode sputtering is the inclusion of the sputter gas in the thin film. Traces of argon can normally be found in argon-sputtered thin films. This can lead to voids in the film which can later get replaced by moisture and results in a slow degradation of the films.

14.2.1.1 DC Sputter Deposition

DC sputter deposition uses a DC power supply to energize the plasma. As a consequence, only electrically conductive materials can be used as targets. The negative terminal of the power supply is attached to the cathode, and the positive terminal is tied to ground. This allows positive gas ions to be drawn toward the cathode and produces sputtering. The cathode terminal is typically much smaller in size than the anode (chamber walls). This size difference will make the electric field strength at the cathode to be much greater than that at the anode. As a result, most of the plasma power will be discharged near the cathode.

It is important to point out that typical DC power supplies such as constant-voltage or constant-current sources cannot be used for energizing a plasma. The $I - -V$ characteristic of a plasma is complex and goes through regions of negative resistances. Therefore, plasma excitation requires a power supply specifically designed for that purpose. Additionally, typical plasma power supplies are used in a constant-power mode rather than a constant-voltage or constant-current mode.

14.2.1.2 RF Sputter Deposition

Deposition of insulating materials requires excitation using an alternating current source. This is most commonly accomplished with an RF source. Due to the potential of interference with radio communications, the FCC has assigned discrete RF frequencies for industrial, scientific, and medical use. One of the most commonly used frequencies for laboratory plasmas is 13.56 MHz. RF plasmas require more complicated setups than DC plasmas. They require an impedance matching network to reduce reflections from the load (in this case, the load is the cathode and the plasma). Despite the added complexity, RF sputtering is widely used because it can be used for sputtering insulators as well as conductors.

Strictly speaking, a symmetric alternating voltage will not have a specific anode or cathode. However, the smaller of the two electrodes in a plasma system will have a higher electric field strength. As a result, ions will impact the smaller electrode with more energy than the larger electrode, resulting in most of the electrons and most of the sputtered atoms emanating from the smaller electrode. Although this is true for DC and RF plasma, in an RF plasma, this asymmetry in electrode geometry leads to a condition where the smaller electrode automatically develops a negative DC bias voltage, as shown in Figure 14.3. As a result, this electrode is also referred to as the cathode, similar to the terminology used in DC sputtering. If the electrodes have equal sizes and are spaced close together such that the effects of the chamber walls are minimal, then the DC bias will be closer to zero, which will result in negligible sputtering. Incidentally, this configuration is used when sputtering is not desired, such as in plasma light sources. Other than the emergence of a DC bias on the RF signal, the deposition process is identical to DC sputtering.

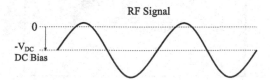

Figure 14.3 DC self-bias on the RF signal.

14.2.1.3 Reactive Sputter Deposition

While the majority of sputtering is done in an argon gas environment, there are instances when a small amount of reactive gas, such as oxygen or nitrogen, is introduced to intentionally cause a reaction between the deposited species and the gas. For example, aluminum oxide (Al_2O_3) can be deposited from a metallic aluminum target with a plasma containing a mixture of argon and oxygen. The sputter yield of aluminum is significantly higher than aluminum oxide, so sputtering pure aluminum and then oxidizing it results in a much faster deposition rate. However, this is not as simple as adding the reactive gas to the mixture. The target will also get oxidized by the reactive gas. This can dramatically alter the deposition rate, a condition known as target poisoning. The goal is to have a net sputter removal rate that is larger than the target oxidation rate, so that the sputtered species consists mostly of the unoxidized target atoms. The reaction with oxygen takes place after they arrive at the substrate surface.

Reactive gases can also be used to compensate for the loss of one species during sputtering. For example, while sputtering Nb_2O_5 with pure argon, the resulting film can be deficient in oxygen because oxygen is more volatile than niobium. By adding a small amount of oxygen to the gas mixture, the loss of oxygen can be compensated.

14.2.1.4 Ion Beam Sputtering

In conventional cathode sputtering, the plasma is generated by the cathode and causes the cathode material (or target) to be sputtered. As illustrated in Figure 14.4, in ion beam sputtering (IBS), the ions are generated from a separate ion gun which is then directed toward the sputter target. This allows a wider range of ion energies and ion fluxes to be used, allowing finer control of the energies of the deposited atoms. Because there is no plasma on the target, inclusion of gas species into the deposited film is also reduced. IBS can be done using both conductive and insulating targets.

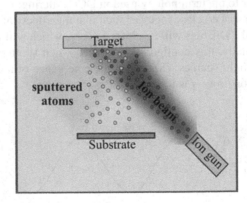

Figure 14.4 Ion beam sputter deposition process.

14.2.2 PLD

PLD is similar to IBS, except the energy for vaporization is delivered via intense laser pulses. The laser pulses ablate the target atoms and eject them into the environment, which are then collected by the substrate at some distance away. Unlike sputtering, plasma is not required. Therefore, the chamber could be under high vacuum or consist of inert or reactive gases. The biggest advantage of PLD compared to cathode sputtering is that the gas environment is nearly independent of the ablation process. In cathode sputtering, the gas environment and the plasma are inherently coupled to each other, which places a limit on the range of pressures that can be used during sputtering. As a result, a broader range of processes can be used with PLD. Similar to sputtering, PLD can be used to deposit alloys and compounds. The laser energy is usually well above the ablation threshold, which results in all atomic components of the target being equally ejected. This allows complex multicomponent compounds to be deposited using PLD compared to other methods. The biggest disadvantage of PLD is the low deposition rate and the poor deposition uniformity. This arises due to the small spot size of the laser and the small volume of ablated material. In order to get a reasonable deposition rate, the substrate has to be placed close to the target, and this results in a gaussian-like distribution of the film thickness. The ejected plume from the target can also contain macroscopic molten material which can get deposited on the substrate as large pieces of debris. These aspects have limited the application of PLD in large production environments (Figure 14.5).

14.2.2.1   Sputter Configurations

Sputter deposition is not influenced by gravity, so it can be configured in many different ways, such as sputter-up, sputter-down, or sputter-sideways. Rotation of the substrate is usually required to make the deposition uniform. While research

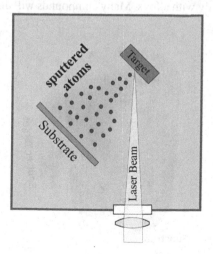

Figure 14.5   PLD.

laboratories primarily use circular targets, their shape in industrial applications greatly depends on the substrate to be coated. Deposition on large sheets use cylindrical or rectangular targets with the substrate rolled on a conveyor system below or above the cathode. Rectangular targets can also be mounted sideways with the substrates mounted along the side walls of a rotating drum. This is shown in Figure 14.6.

14.2.3 THERMAL EVAPORATION

Every material is constantly evaporating, even at room temperature. The vapor pressure of the material is the primary factor that determines its evaporation rate. Raising the temperature will dramatically increase the vapor pressure and will raise the evaporation rate. This is the principle behind thermal evaporation. In thermal evaporation, a source material is heated until its vapor pressure becomes high enough to produce a reasonable vaporization rate. The substrate is placed in the path of the vapor stream to cause it to condense as a thin film. It is conceptually very simple and easy to implement. Metals typically have to be raised above their melting temperature to cause an appreciable increase in vapor pressure. This can range from about 1,000°C for aluminum to about 2,500°C for platinum. Some metals, such as chromium, evaporate without melting, which is known as sublimation. Dielectrics behave the same way, and many of them have to be raised to temperatures in the range of 2,000°C to evaporate. The characteristic difference of evaporation compared to sputtering is the absence of a plasma. This also means that the background pressure during deposition will be much lower, usually in the range of μTorr. The deposition energies are also much lower in evaporation. Despite the high temperatures, the arrival energy of the vapor molecules is usually <1 eV (whereas in sputtering, it is in the range of 10 eV).

A major disadvantage of thermal evaporation is the difficulty in maintaining stoichiometry, especially with alloys. Many compounds will disintegrate when raised

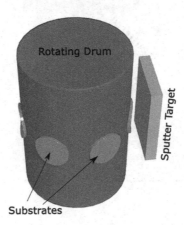

Figure 14.6 Drum configuration.

to high temperatures, and each component of the mixture will evaporate separately at different rates leading to a completely different stoichiometry at the substrate surface. As a result, only the most stable compounds can be deposited using thermal evaporation. Fortunately, most oxides and nitrides used in optical thin films are stable enough to be deposited using evaporation. These include SiO_2, MgF_2, Al_2O_3, Ta_2O_5, and Si_3N_4. Alloys such as ITO and GST cannot be easily deposited by thermal evaporation because their components will evaporate separately at different rates. Flash evaporation is a type of evaporation that can be used to evaporate components with different vapor pressures. The source material, usually in powder form, is dispensed at a controlled rate into a heated crucible to produce rapid evaporation. Since the temperature of the crucible is usually much higher than that required for normal evaporation, all species evaporate rapidly at similar rates (similar to PLD).

Types of thermal evaporation differ primarily in how the heat is delivered to the source material. The two primary techniques are resistive heating and electron-beam heating.

14.2.3.1 Resistively Heated Thermal Evaporation

Thermal energy for evaporation can be delivered by simple ohmic heating by electric current. As illustrated in Figure 14.7, a large DC current is flowed through a tungsten or molybdenum foil that is shaped like a boat or dish. The material to be evaporated is placed in this boat and heated. This is the simplest and most inexpensive evaporation technique. The disadvantage of this method is the high electrical current required to

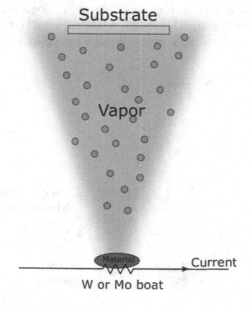

Figure 14.7 Resistively heated thermal evaporation.

reach the temperatures necessary for evaporation. Collateral heating is a common problem, which can cause evaporation from unwanted surfaces leading to contamination of the evaporating species.

14.2.3.2 Flash Evaporation

In flash evaporation, the crucible is maintained at a temperature higher than that required for normal evaporation. The material to be deposited is fed into the crucible at a constant rate. The goal is to evaporate the material rapidly at a temperature well above its evaporating threshold such that all components will evaporate fully. In other words, the evaporation rate has to be equal to the feed rate of the material, which will ensure that the deposited film will have the same stoichiometry as the source material.

14.2.3.3 Electron-Beam-Heated Thermal Evaporation

As shown in Figure 14.8, in electron-beam evaporation, thermal energy is delivered to the source via energetic electrons accelerated in vacuum. In a typical configuration, free electrons are produced by thermionic emission from a hot tungsten filament. The electrons are then accelerated by several thousands of volts, and magnetic fields are used to redirect and focus the beam of electrons toward the source material. Upon colliding with the source material, the electrons will quickly lose their excess energy within a few collisions, causing a large temperature rise within a short distance. It results in the exterior surface of the material being heated much more than the interior. Therefore, compared to heating the entire material volume, electron beam heating is able to produce high vapor pressures using a lower electrical power compared to resistive heating. This results in less collateral heating and less contamination. However, it comes at the cost of greater equipment complexity. It requires

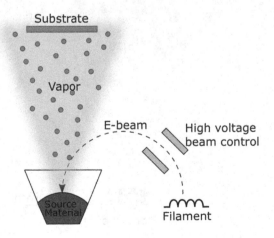

Figure 14.8   Electron-beam thermal evaporation process.

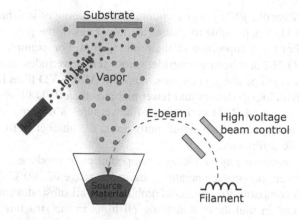

Figure 14.9 Ion-assisted electron beam evaporation.

circuits to control and direct the electron beam, as well as high-voltage power supplies and the associated safety issues.

14.2.3.4 Reactive Evaporation

Just like in sputtering, reactive mechanisms can be employed during evaporation. Oxygen or nitrogen can be introduced to create oxides or nitrides. However, excessive reaction can also degrade the electron beam source (usually a tungsten filament). Grounding plates used for providing a path for the electrons to reach ground can also get oxidized and cause charge to build up, which will deflect the electron beam. Additionally, excessive background pressure can cause the electron beam to scatter and even lead to arcing.

14.2.3.5 Ion-Assisted Deposition

Ion-assisted deposition (IAD) combines normal evaporation or sputtering with an ion beam directed at the substrate. It is similar to IBS, except the ion beam is aimed at the substrate rather than the target. This is shown in Figure 14.9. The ions bombard the substrate during deposition densifying the film. The ions can be from an inert gas such as argon, or they can also be used with a reactive gas such as oxygen or nitrogen.

14.3 CHEMICAL VAPOR DEPOSITION

As the name implies, chemical vapor deposition (CVD) utilizes vapor precursors to produce a thin film through a chemical reaction. For example, Si_3N_4 can be produced by reacting SiH_4 (silane gas) with NH_3 (ammonia gas). The reaction rate will be primarily determined by the substrate temperature and partial pressures of the precursors. An important advantage of CVD is the ability to fine-tune the stoichiometry of the film. In PVD, the film and the source material will have nearly the same stoichiometry. Reactive PVD is an exception, but fine-tuning of stoichiometry is still not easy. In CVD, it is possible to change the flow rates of the precursor gases and significantly alter the composition of the resulting film. For example, it is possible to produce $Si_xO_yN_z$, or silicon oxynitride, whose refractive index can be tuned between 1.45 and 2.0 depending on the stoichiometry. Also, CVD films have generally better quality with higher density and fewer pinholes than PVD films. However, the biggest disadvantage of CVD is that each film requires a different chemistry and different precursor gases, which significantly limits the number of films that can be grown in a single equipment.

Most CVD reactions require elevated temperatures to produce a reasonable reaction rate. Typical process temperatures are in the range of 300°C–1,000°C. This makes material compatibility a potential problem. Not all substrates can survive high temperatures, and in multilayer thin films, all films in the structure must be able to withstand the deposition temperatures. Although there are some low-temperature CVD processes, they come at the expense of poorer film quality. On the other hand, in PVD, heating the substrate is optional and is not a fundamental requirement.

Another important characteristic of CVD is the conformal nature of the deposition. As mentioned previously, PVD is highly directional, which results in deposition rates that depend on substrate position and angle. CVD, on the other hand, is highly conformal and uniform. It is ideally suited for complex shapes where all surfaces need to be coated equally. As a result, substrate rotation is rarely necessary in CVD.

There are many different types of CVD methods, and not all of them are relevant for optical thin film applications. Deposition methods that are used in optical thin films include low-pressure chemical vapor deposition (LPCVD), plasma-enhanced chemical vapor deposition (PECVD), and atomic layer deposition (ALD).

14.3.1 LPCVD

LPCVD is done at pressures in the range of 1–10 Torrs. In PVD, these would be considered high pressures, but in CVD, these are actually considered low pressures. This is because CVD reaction rates are directly related to pressure, so the reactions must be conducted at much higher pressures. The temperatures used in LPCVD are in the range of 700°C–1,000°C. The most commonly used LPCVD films are silicon nitride, silicon dioxide, and amorphous silicon. Silane, ammonia, and nitrous oxide are some of the precursor gases used in these processes. Nevertheless, the high substrate temperatures required for LPCVD rule out many multilayer configurations, and it is used only when all other options are not possible.

14.3.2 PECVD

PECVD is a variant of LPCVD where a plasma is introduced near the substrate to replace the high temperatures in LPCVD. This allows reactions to occur at lower temperatures around 300°C. The objective here is temperature reduction, not ion

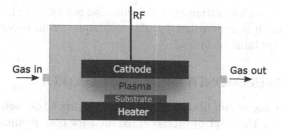

Figure 14.10 PECVD.

bombardment. Therefore, the electrodes are designed to be equal in size, which results in a nearly zero DC bias with an RF excitation. PECVD was developed primarily by CMOS fabrication, but it is also suitable for optical thin films such as silicon nitride and silicon dioxide. The plasma also introduces some directionality to the process. As a result, PECVD thin films will not be completely conformal, whereas LPCVD thin films are. A typical PECVD system is illustrated in Figure 14.10.

14.3.3 ALD

ALD is a CVD process that has become popular in recent years. Whereas in LPCVD and PECVD the deposition runs in one long continuous cycle, in ALD the deposition takes place in short cycles. Each cycle results in a self-terminating reaction, so the reaction rate and the resulting film thickness can be controlled very precisely.

The gas precursors in ALD are flowed one at a time as gas pulses. The process relies on gas molecules that get adsorbed on the substrate surface rather than the bulk content. For example, for depositing Al_2O_3, the precursor gases are $C_6H_{18}Al_2$ (trimethylaluminum) and H_2O. During the first pulse, trimethylaluminum is flowed and allowed to adsorb to the surfaces. Then argon is used to flush out the gas from the chamber. Then water vapor is flowed in. This reacts with the adsorbed trimethylaluminum molecules from the first pulse, producing a very thin layer of Al_2O_3. Then argon is flowed again to flush out the gases, which completes one cycle. The film thickness achieved during each cycle will be determined by the surface properties and temperature and not by the duration of the gas pulses. The reaction will come to a stop when all adsorbed molecules are depleted. This allows us to count the cycles to accurately control the film thickness. ALD can also be combined with plasma activation to improve the reaction rates. To protect the substrate from ion bombardment, typical configurations use a downstream plasma. In this configuration, the plasma is generated upstream in a separate chamber, and the resulting gases containing the free radicals are collected and flowed through the ALD reaction chamber. A major disadvantage of ALD is the very small film thickness that is produced during each cycle. The thickness per cycle is typically in the range of 1–10 . Depositing a film of 100 nm thickness would take several hours. When combined with the fact that each film type would require at least two different precursor gases, ALD becomes less attractive for conventional optical thin film applications. However, they do have many other advantages, such as extremely conformal and pinhole-free films for ultrathin-film applications. It is anticipated that this technique would become more widely used in the future (Table 14.1).

14.4 THICKNESS MONITORING AND CONTROL

Optical performance of thin-film structures critically depend on their thicknesses and refractive indices. The level of precision required for optical films is significantly greater than that required in other areas of thin-film applications. Even a 2 nm deviation of a high-index film will cause a noticeable spectral distortion. Several techniques are employed for monitoring the thin-film deposition during growth. Some of these are discussed next.

14.4.1 QUARTZ CRYSTAL MICROBALANCE

Quartz crystal microbalance (QCM) is the most widely used technique for real-time monitoring of thin films. Although it is not particularly suited for optical thin films, it warrants discussion anyway due to its ubiquitous use in nearly every thin-film-deposition system.

The QCM can be thought of as an extremely sensitive miniature weighing scale. It consists of a thin crystal of piezoelectric quartz. Acoustic waves are induced inside the crystal by electrodes on either side of the crystal faces. This sets up a stand-ing wave resonance. The resonance frequency is a function of the crystal thickness. Typical crystals are designed to be resonant at 6 MHz. Any material that adheres to the surface will change the resonance frequency, and this can be viewed as a mass attached to the end of an oscillating spring. The resonant frequency drops as more and more material builds up on the crystal faces. By measuring the resonance fre-quency, information about the deposition can be deduced. An illustration of a QCM is shown in Figure 14.11. The equation that describes this relationship is known as the Sauerbrey equation, given by

$$\frac{\Delta f}{f} = \frac{2f}{z_q}m, \qquad (14.1)$$

where f is the resonance frequency, Δf is the change in resonant frequency, z_q is the acoustic impedance of quartz, and m is the mass per unit area on the QCM surface. By measuring the frequency shift Δf, we can calculate m, which can then be divided by the material density to get film thickness.

Equation (14.1) treats the film as a point source mass attached to the end of the quartz substrate. The acoustic impedance and acoustic thickness of the film are ig-nored. As a result, this model becomes less and less accurate as the film becomes thicker. This can be corrected to some extent by treating the film similar to how we treat the optical effects of thin films on a substrate. In other words, we can include the effects of the acoustic impedance and acoustic thickness of each film in the model. This is known as the Z-match method and is currently the most widely used model in thin-film measurements. Nevertheless, for accurate results, one has to know the

Table 14.1

Table Summary of Deposition Methods and Their Common Characteristics

Method/Variants	Action	Materials	Pros	Cons	Options
Sputtering					
–RF	Cathode ion bombardment	All materials with sufficiently high sputter yield	Flexible geometries; compounds and alloys can be deposited	Complex setup (RF generator, impedance matching)	Inert or reactive
–DC	Cathode ion bombardment	Conductive	Simpler setup	High energy	Inert or reactive
–Ion beam	Ion bombardment from a separate gun	All materials with sufficiently high sputter yield			Inert or reactive
PLD	Laser ablation	All	No gas required	Poor uniformity, low rates	Inert or reactive
Evaporation					
–Resistive	Ohmic heating	All materials with sufficiently high vapor pressure	Simpler setup	Contamination from collateral heating	Ion assist, flash, inert, or reactive
–E-beam	High energy electrons	All materials with sufficiently high vapor pressure	High purity	Complex setup, high voltage hazards	Ion-assist, flash, inert, or reactive
CVD					
–LPCVD	Thermally activated chemical reaction	Depends on available precursors	Robust films, stoichiometry control, conformal	High temperature	
–PECVD	Plasma-activated chemical reaction	Depends on available precursors	Relatively lower temperatures, stoichiometry control	More complex setup	
–ALD	Monolayer chemical reaction	Depends on available precursors	Simple setup, self-terminating, conformal	Slow deposition rates	Plasma assist

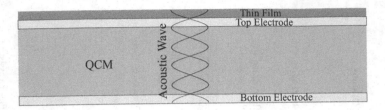

Figure 14.11   QCM.

density and acoustic impedance of the film. These parameters may not be accurately known for every material. Additionally, the film thickness on the QCM will not be the same as the film thickness on the substrate due to differences in geometric placement between the substrate and QCM. Also, the sensitivity of the QCM declines depending on the penetration of the acoustic waves into that film. This in turn will depend on the layer stack. In general, however, the sensitivity of the QCM will decline as more and more film accumulates on the crystal face, which makes it less accurate for thick films. Since most optical structures utilize many layers that can add up to several micrometers in total thickness, this can become a serious limitation. Most importantly, what we care about in optical thin-film designs are the refractive indices and optical phase thicknesses of the films. QCM, on the other hand, measures the mass and acoustic impedances of the films. So one has to rely on previously known relationships between these measured quantities and their optical constants. In fact, even the optical constants of the films are not as important as their overall optical performance (reflection and transmission). For all of these reasons, despite the widespread use of QCM in most areas of thin films, it is not a very reliable method for optical thin film deposition.

14.4.2   OPTICAL MONITORING

Direct optical monitoring of the films during deposition can yield very useful information about the refractive indices and film thicknesses. With suitable view ports on the vacuum chamber, it is possible to place a laser and a photodetector to measure the optical characteristics of the films during deposition. While some of the early implementations used a laser and a photodetector, with the availability of compact non-scanning (CCD (charge coupled device)) spectrometers, it has become commonplace to use a broadband light source and interrogate the entire reflection (or transmission) spectrum of the sample during deposition. Such measurements allow immediate and direct evaluation of the thin-film performance and make real-time corrections to the deposition process.

Many variations of this concept are used in the field, and one such example is shown in Figure 14.12. This setup utilizes a fiber optic spectrometer and a broadband light source such as a tungsten–halogen lamp. The light is delivered to the sample through a fiber bundle. A collimator is useful for reducing the losses due to beam divergence. Light is incident on the backside of the substrate through a suitable view

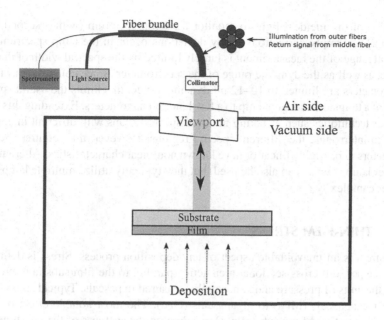

Figure 14.12 Optical monitoring setup with a spectrometer.

port, and the reflected signal is collected by the same collimator and sent through one of the fibers in the bundle to the spectrometer. With fast CCD-based spectrometers, it is possible to monitor the reflectivity *in-situ* without retrieving the sample from the vacuum chamber.

However, there are a number of practical considerations to such a system. The deposition process will create a significant amount of background light. Thermal evaporation produces copious amounts of radiation, and plasmas used in sputtering contain strong optical emission lines. As a result, the measurement will contain large portions of this radiation and will significantly reduce the signal-to-noise ratio. A commonly used technique to improve the signal is by modulating the light source and demodulating the signal from the spectrometer. A simpler solution is to momentarily halt the deposition while taking the measurement, but this would compromise the ability to acquire real-time measurements. Reflections from the viewports as well as unintentional depositions accumulating on the viewport windows must also be considered. Wedged viewport windows are commonly employed to eliminate these effects. A shutter may have to be used to block deposition species coating the viewport windows. The stability of the lamp is also an important consideration. At the beginning of the deposition, the lamp is calibrated by applying the known reflection from the substrate, which is subsequently used to determine the spectral reflection or transmission from the films. Any intensity or spectral drifts of the lamp will alter this baseline data and will produce errors in the subsequent measurements. Therefore, a highly stable light source is essential for acquiring reliable data. Some systems use a portion of the incident light to monitor the lamp spectrum (with a second spectrometer) and apply corrections if any deviations occur in the lamp spectrum. The spectral range of the measurement is largely limited by the spectral width of the light source, as well as the dynamic range of the spectrometer. Since commonly available spectrometers are limited to 10–12 bit dynamic range, this limits the useful spectral range of a tungsten–halogen lamp to a few hundred nanometers. Extending this range requires techniques such as taking multiple spectral scans with different integration times to interrogate the different spectral regions. However, this requires the photodetectors to be highly linear or have known nonlinear characteristics. Alternatively, broader lamp sources can also be used, but they typically utilize multiple lamps with greater complexity.

14.5 THIN-FILM STRESS

Film stress is an unavoidable aspect of any deposition process. Stress is defined as the force per unit cross-sectional area acting parallel to the film/substrate interface. It has the units of pressure and is normally measured in pascals. Typical stress values are in megapascals (MPa) to gigapascals (GPa). The most notable effect of stress is the deformation of the substrate. Depending on the stiffness of the substrate (i.e., Young's modulus E), this will produce a curvature in the substrate.

Stress is categorized as compressive or tensile. A compressively stressed film is under too much compression and will exhibit a tendency to expand in order to reduce its internal energy. This will make the substrate curve outward and away from the film. A tensile film is under too much tension and has to contract in order to reduce its internal energy. This will make the substrate curve inward. Both of these assume that the adhesion between the film and the substrate is sufficiently high. If the adhesion is weak, a stressed film will simply delaminate from the substrate instead of bending the substrate. Due to the profound effects of stress, quantifying and managing film stress is an important aspect of any thin-film study (Figure 14.13).

There are two primary sources of stress – extrinsic and intrinsic. Extrinsic stress is the simplest to understand. It arises due to external factors such as temperature and moisture. When the temperature increases, each film and substrate will expand at different rates depending on their coefficients of thermal expansion (CTEs). A large mismatch in CTE values will lead to some compressive or tensile stress in the films.

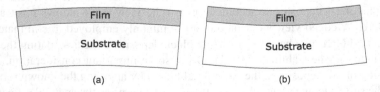

          (a)   (b)

Figure 14.13 Effect of a single film with compressive stress or tensile stress on the substrate curvature. The curvature is highly exaggerated in this figure for illustration. (a) Compressive stress. (b) Tensile stress.

Penetration of moisture in the films can also cause a similar effect. If the devices will be exposed to large temperature swings and/or moisture, this is an important aspect that should be considered. The effects can be mitigated by choosing materials with similar CTE as well as passivation coatings to prevent the penetration of moisture.

The largest source of thin-film stress, however, is intrinsic. This is dictated by the growth mechanism. Sputtered films generally have compressive characteristics because the incident energies of the sputtered atoms is quite large (several electron-volts). The high arrival velocities result in the films being highly compacted and compressed. Evaporated films display the opposite behavior – they are generally tensile because the arrival energies are very small (several hundred millielectronvolts). CVD films can display either characteristic depending on the reaction mechanism. Reactions that are dominated by inclusions will produce compressive films, and those dominated by desorption can lead to tensile films. However, these are general guidelines and should not be taken as always true. Some sputtered films can be tensile, and some evaporated films can be under compression. Therefore, it is important to accurately quantify the stress values.

Stress is the force imparted by the film on the substrate. This can be indirectly deduced by measuring the curvature of the substrate induced by the film. The tangential force per unit area required to produce a deflection on a substrate whose thickness is t_s can be related to its Young's modulus E (elasticity) and Poisson's ratio (change in dimension in the transverse direction) as

$$\sigma = \frac{E_s}{6(1-v_s)} \frac{t_s^2}{t_f} \left(\frac{1}{r_2} - \frac{1}{r_1} \right), \tag{14.2}$$

where r_1 is the original radius of curvature of the substrate and r_2 is the radius of curvature after the film.

FURTHER READING

Badoil, B., Lemarchand, F., Cathelinaud, M. & Lequime, M. Interest of broadband optical monitoring for thin-film filter manufacturing. *Applied Optics* **46**, 4294–4303 (July 2007).

Jilani, A., Abdel-wahab, M. S. & Hammad, A. H. *Modern Technologies for Creating the Thin-film Systems and Coatings* (InTech Ltd, Croatia, 2017).

Lakhtakia, A. & Messier, R. *Sculptured Thin Films: Nanoengineered Morphology and Optics* (SPIE Publications, Bellingham, WA, 2005).

Macleod, H. A. Monitoring of optical coatings. *Applied Optics* **20**, 82–89 (January 1981).

Macleod, H. A. *Optical Thin Films and Coatings* (eds Piegari, A. & Flory, F.) 3–25 (Woodhead Publishing, Sawston, 2013). ISBN: 978-0-85709-594-7.

Mahan, J. E. *Physical Vapor Deposition of Thin Films* (Wiley- Interscience, NY, 2000). ISBN: 0471330019.

Sarangan, A. *Nanofabrication: Principles to Laboratory Practice.* (CRC Press, Boca Raton, FL, 2016). ISBN: 9781498725576.

15 Python Computer Code

15.1 PLANE WAVE TRANSFER MATRIX METHOD

15.1.1 SUBROUTINE: TMM.PY

```python
import numpy as np

### Subroutine for forming the 2x2 matrix
def tmm_matrix(nf,z,k0,q,na,sp):
    nx = na*np.sin(q)
    nz = np.sqrt(nf**2-nx**2);
    if sp == ''TE'': #TE/TM polarization
        m11 = np.exp(-1j*k0*nz*z)
        m12 = np.exp(1j*k0*nz*z)
        m21 = nz*np.exp(-1j*k0*nz*z)
        m22 = -nz*np.exp(1j*k0*nz*z)
        M = np.array([[m11,m12],[m21,m22]])
    if sp == ''TM'':
        m11 = nz/nf*np.exp(-1j*k0*nz*z)
        m12 = -nz/nf*np.exp(1j*k0*nz*z)
        m21 = nf*np.exp(-1j*k0*nz*z)
        m22 = nf*np.exp(1j*k0*nz*z)
        M = np.array([[m11,m12],[m21,m22]])
    return M

def tmm_n(nf,z1,z2,k0,q,na,sp):
    M1 = tmm_matrix(nf,z1,k0,q,na,sp)
    M2 = tmm_matrix(nf,z2,k0,q,na,sp)
    N = np.matmul(M2,np.linalg.inv(M1))
    return N


def tmm(wavelength,ns,ts,na,nf,thk,q,sp,back):
    nsi = np.imag(ns)
    tz = np.cumsum(np.insert(np.multiply(thk,-1.0),0,0.0))
    nwp = wavelength.size    #number of wavelength points
    nl = nf.size/nwp         #number of layers
    k0 = 2.0*np.pi/wavelength
    Rr = np.zeros(nwp)
    Rt = np.zeros(nwp)
    Tr = np.zeros(nwp)
    Tt = np.zeros(nwp)
    T = np.zeros(nwp)
    R = np.zeros(nwp)
    t = np.zeros(nwp,dtype=complex)
    r = np.zeros(nwp,dtype=complex)

    for i in np.arange(0,nwp,1):        #Step through all wavelengths
        Ms = tmm_matrix(ns[i],tz[0],k0[i],q,na,sp)
```

```
N = np.array([[1.0,0.0],[0.0,1.0]])
for n in np.arange(0,nl,1):    #Step through all layers
    N = np.matmul(tmm_n(nf[n*nwp+i],tz[n],tz[n+1],k0[i],
        q,na,sp),N)
Ma = np.linalg.inv(tmm_matrix(na,tz[nl],k0[i],q,na,sp))
M = np.matmul(N,Ms)
M = np.matmul(Ma,M)
r[i] = M[1,0] / M[0,0]
R[i] = np.absolute(r[i])**2
t[i] = 1.0/M[0,0]
T[i] = np.absolute(t[i])**2

if back == 1: #If back=1, back interface will be computed.
  Otherwise ignore.
    Rs = np.absolute((ns[i]-na)/(ns[i]+na))**2
    Ts = np.absolute((1.0-Rs)*ns[i]/na)
    Rr[i] = np.absolute(M[0,1]/M[0,0])**2
    Tr[i] = np.absolute(M[1,1] - M[0,1]*M[1,0]/M[0,0])**2
    Rt[i] = R[i] + Tr[i]*Rs*T[i]*np.exp(4.0*k0[i]*nsi[i]
        *ts*1000.0)/\
        (1.0-Rr[i]*Rs*np.exp(4.0*k0[i]*nsi[i]*ts*1000.0))
    Tt[i] = Ts*T[i]*np.exp(2.0*k0[i]*nsi[i]*ts*1000.0)/\
        (1.0-Rr[i]*Rs*np.exp(4.0*k0[i]*nsi[i]*ts*1000.0))
else:
    Rt[i] = R[i]
    Tt[i] = T[i]*np.absolute(ns[i]/na)

return [Tt,Rt,t,r]
```

15.1.2   TMM REFLECTION SPECTRUM WITH AND WITHOUT SUBSTRATE BACKSIDE (FIGURE 4.4 IN CHAPTER 4)

```
#!/usr/bin/python

import numpy as np
from tmm import tmm
import matplotlib.pyplot as plt

nlambda = 1000 #Number of wavelength points
wavelengths=np.linspace(400,800,nlambda)#Wavelength array 400nm-800nm
ns = np.linspace(1.48,1.48,nlambda) #Substrate index array
ts = 500  #Substrate thickness- not relevant unless back=1
na = 1.0  #Incident medium (air) index
nf = np.linspace(1.217,1.217,nlambda) #Index of layers
thk = np.array([550/4/1.217])  #thickness of layers
q = 0.0*np.pi/180.0  #Incident angle
sp = 'TM'  #Incident polarization (only for off-normal incidence)
back = 0  #back=0 will ignore backside of the substrate
[T,R,t,r] = tmm(wavelengths,ns,ts,na,nf,thk,q,sp,back)    #Run TMM
back = 1  #back=1 will include backside of the substrate
[T,Rb,t,r] = tmm(wavelengths,ns,ts,na,nf,thk,q,sp,back)    #Run TMM

np.savetxt('tmm_R_withwithout.data', np.transpose([wavelengths,R,Rb]))
```

```
   #save to file
plt.plot(wavelengths,R,wavelengths,Rb)
plt.xlabel('Wavelength')
plt.ylabel('Reflection')
plt.show()
```

15.1.3 TMM REFLECTION SPECTRUM INCLUDING COMPLEX AND DISPERSIVE MATERIALS (FIGURE 4.5 IN CHAPTER 4)

```
#!/usr/bin/python

import numpy as np
from tmm import tmm
import matplotlib.pyplot as plt

#The external file contains a three column data of
#wavelength (in um), Real Index and Imag Index
arr = []
f = open('Si.data', 'r')
for line in f:
    arr.append(map(float, line.split()))
r1 = zip(*arr)
f.close

arr = []
f = open('SiO2.data', 'r')
for line in f:
    arr.append(map(float, line.split()))
r2 = zip(*arr)
f.close

arr = []
f = open('Cu.data', 'r')
for line in f:
    arr.append(map(float, line.split()))
r3 = zip(*arr)
f.close

nlambda = 1000
wavelengths=np.linspace(400,1000,nlambda)
nsi =  np.interp(wavelengths,np.asarray(r1[0])*1000.0,\
        np.asarray(r1[1])-1j*np.asarray(r1[2]))
nsio2 =  np.interp(wavelengths,np.asarray(r2[0])*1000.0,\
        np.asarray(r2[1])-1j*np.asarray(r2[2]))
ncu =  np.interp(wavelengths,np.asarray(r3[0])*1000.0,\
        np.asarray(r3[1])-1j*np.asarray(r3[2]))
ns = nsi
nf = np.concatenate((nsio2,nsi,ncu))
thk = np.array([240.0,10.0,5.0])  #thickness of layers

ts = 500   #Substrate thickness- not relevant unless back=1
na = 1.0   #Incident medium (air) index
q = 0.0    #Incident angle
```

```
sp = 'TM'  #Incident polarization (only for off-normal incidence)
back = 0   #back=0 will ignore backside of the substrate
[T,R,t,r] = tmm(wavelengths,ns,ts,na,nf,thk,q,sp,back)       #Run TMM

np.savetxt('tmm_dispersion_complex.data',
 np.transpose([wavelengths,R])) #save to file
plt.plot(wavelengths,R)


plt.xlabel('Wavelength')
plt.ylabel('Reflection')
plt.show()
```

15.1.4   SUBROUTINE: TMM_FIELD.PY

```
import numpy as np
from tmm import tmm_matrix

def tmm_field(wavelength,zp,ta,na,ts,ns,nf,thk,q,sp,r):

    tz = np.cumsum(np.insert(np.multiply(thk,-1.0),0,0.0))

    #Field in the incident medium
    zd = np.linspace(tz[-1]-ta,tz[-1],zp)
    index = np.linspace(na,na,zp)
    C = np.array([[1],[r[0]]])
    k0 = 2.0*np.pi/wavelength[0]
    kz = k0*np.sqrt(na**2-(na*np.sin(q))**2)
    Fz = C[0][0]*np.exp(-1j*kz*zd) + C[1][0]*np.exp(+1j*kz*zd)
    n1 = nf.size
    n2 = na

    #Field in the films
    for n in np.arange(n1-1,-1,-1):          n1 = n2
        n2 = nf[n]
        M1 = tmm_matrix(n1,tz[n+1],k0,q,na,sp)
        M2 = np.linalg.inv(tmm_matrix(n2,tz[n+1],k0,q,na,sp))
        M = np.matmul(M2,M1)
        zdn = np.linspace(tz[n+1],tz[n],zp)
        C = np.dot(M,C)
        kz = k0*np.sqrt(n2**2-(na*np.sin(q))**2)
        Fzn = C[0][0]*np.exp(-1j*kz*zdn) + C[1][0]*np.exp(+1j*kz*zdn)
        zd = np.append(zd,zdn)
        idx = np.linspace(n2,n2,zp)
        index = np.append(index,idx)
        Fz = np.append(Fz,Fzn)

    #Field in substrate
    zdn = np.linspace(tz[0],tz[0]+ts,zp)
    idx = np.linspace(ns[0],ns[0],zp)
    n1 = n2
    n2 = ns[0]
    M1 = tmm_matrix(n1,tz[0],k0,q,na,sp)
    M2 = np.linalg.inv(tmm_matrix(n2,tz[0],k0,q,na,sp))
```

```
M = np.matmul(M2,M1)
C = np.dot(M,C)
kz = k0*np.sqrt(n2**2-(na*np.sin(q))**2)
Fzn = C[0][0]*np.exp(-1j*kz*zdn) - C[1][0]*np.exp(+1j*kz*zdn)
zd = np.append(zd,zdn)
index = np.append(index,idx)
Fz = np.append(Fz,Fzn)


return [zd,Fz,index]
```

15.1.5   FIELD PROFILE INSIDE SINGLE-LAYER ANTIREFLECTION (FIGURE 4.6 IN CHAPTER 4)

```
#!/usr/bin/python


import numpy as np
from tmm import tmm
from tmm_field import tmm_field
import matplotlib.pyplot as plt

wavelength=np.array([550.0])   #Wavelength (specify only one)
ts = 500                       #Substrate thickness
ns = np.array([1.48])          #Substrate index
na = 1.0                       #Incident medium (air) index
nf = np.array([1.217])         #Index of layer 1
thk = np.array([550/4/1.217])  #thickness of each layer
q = 0.0*np.pi/180.0            #Incident angle
sp = 'TE'                      #Incident polarization
                               (only for off-normal incidence)
back = 0                       #back=0 will ignore backside
[T,R,t,r] = tmm(wavelength,ns,ts,na,nf,thk,q,sp,back) #Run TMM

ta = 500 #thickness of the air layer
zp = 100 #number of grid points in each layer

[zd,Fz,index] = tmm_field(wavelength,zp,ta,na,ts,ns,nf,thk,q,sp,r)

np.savetxt('tmm_field_slqw_550.data',
 np.transpose([zd,np.absolute(Fz),index])) #save to file
plt.plot(zd,np.absolute(Fz),zd,index)
plt.xlabel('Z (um)')
plt.ylabel('Field & Index')
plt.show()
```

15.1.6   COUPLED-CAVITY LINE FILTER (FIGURE 9.14 IN CHAPTER 9)

```
#!/usr/bin/python


import numpy as np
from tmm import tmm
import matplotlib.pyplot as plt


nlambda = 3000
```

```python
wavelengths=np.linspace(500,600,nlambda)
ns = np.linspace(1.0,1.0,nlambda)
ts = 500  #Substrate thickness- not relevant unless back=1
na = 1.0  #Incident medium (air) index
q = 0.0   #Incident angle
sp = 'TM' #Incident polarization (only for off-normal incidence)
back = 0  #back=0 will ignore backside of the substrate
wl = 550.0

nH = 2.5
nL = 1.5
nfH = np.linspace(nH,nH,nlambda)
nfL = np.linspace(nL,nL,nlambda)
tH = wl/nH/4.0
tL = wl/nL/4.0

nf = np.concatenate((\
        nfL,\
        nfL,nfH,nfL,nfH,nfL,\
        nfL,\
        nfL,nfH,nfL,nfH,nfL,nfH,nfL,nfH,nfL,\
        nfL,\
        nfL,nfH,nfL,nfH,nfL,nfH,nfL,nfH,nfL,nfH,nfL,\
        nfL,\
        nfL,nfH,nfL,nfH,nfL,nfH,nfL,nfH,nfL,nfH,nfL,\
        nfL,\
        nfL,nfH,nfL,nfH,nfL,nfH,nfL,nfH,nfL,\
        nfL,\
        nfL,nfH,nfL,nfH,nfL,\
        nfL))
thk = np.array([\
        tL/2.0,\
        tL/2.0,tH,tL,tH,tL/2.0,\
        tL,\
        tL/2.0,tH,tL,tH,tL,tH,tL,tH,tL/2.0,\
        tL,\
        tL/2.0,tH,tL,tH,tL,tH,tL,tH,tL,tH,tL/2.0,\
        tL,\
        tL/2.0,tH,tL,tH,tL,tH,tL,tH,tL,tH,tL/2.0,\
        tL,\
        tL/2.0,tH,tL,tH,tL,tH,tL,tH,tL/2.0,\
        tL,\
        tL/2.0,tH,tL,tH,tL/2.0,\
        tL/2.0,\
        ])
[T,R,t,r] = tmm(wavelengths,ns,ts,na,nf,thk,q,sp,back)      #Run TMM

np.savetxt('tmm_R245_5coupled.data', np.transpose([wavelengths,R]))
 #save to file
plt.plot(wavelengths,R)
plt.xlabel('Wavelength')
plt.ylabel('Reflection')
plt.show()
```

15.2   EFFECTIVE REFLECTANCE INDEX CONTOURS

15.2.1   SUBROUTINE: CONTOUR.PY

```python
import numpy as np
#ns = substrate (or equivalent)
#nf = film index
#phase = in units of pi/2 (real part only)
#wl = wavelength in nm
#npts = number of points in arc
def contour(ns,nf,phase,wl,npts):
    phase = phase*(1.0+ 1j*np.imag(nf)/np.real(nf))
    q = np.linspace(0,phase*np.pi/2,npts)
    nr =  nf * ((ns + nf) + (ns - nf)*np.exp(-1j*2*q)) \
            / ((ns + nf) - (ns - nf)*np.exp(-1j*2*q))
    cr = np.real(nr)
    ci = np.imag(nr)
    qr = np.real(q)
    qi = np.imag(q)
    t = np.real(q/(2.0*np.pi/wl)/nf)
    return cr,ci,t,qr,qi
```

15.2.2   SINGLE QUARTER-WAVE CONTOUR (FIGURE 3.3 IN CHAPTER 3)

```python
#!/usr/bin/python

import numpy as np
import matplotlib.pyplot as plt
from contour import contour

ns = 1.48    #Substrate index
nf = 1.217   #Film index
phase = 1.0 #Phase in units of pi/2 (QW)
npts = 1000 #Number of points in the film
wl = 550.0 #Wavelength

(cr,ci,t,qr,qi) = contour(ns,nf,phase,wl,npts)

np.savetxt('contour_slqw.data', np.transpose([cr,ci])) #save to file
plt.plot(cr,ci)
plt.xlabel('Real Index')
plt.ylabel('Imaginary Index')
plt.show()
```

15.2.3   TWO QUARTER-WAVE CONTOURS (FIGURE 5.1 IN CHAPTER 5)

```python
#!/usr/bin/python

import numpy as np
import matplotlib.pyplot as plt
from contour import contour

ns = 1.48
```

```
npts = 1000                    #Number of points in the film

cr = np.zeros(shape=(2,npts))
ci = np.zeros(shape=(2,npts))
t = np.zeros(shape=(2,npts))
qr = np.zeros(shape=(2,npts))
qi = np.zeros(shape=(2,npts))

nf = 1.68
phase = 1.0
(cr[0],ci[0],t[0],qr[0],qi[0]) = contour(ns,nf,phase,550.0,npts)
ns = cr[0][-1]+1j*ci[0][-1]
nf = 1.38
(cr[1],ci[1],t[1],qr[1],qi[1]) = contour(ns,nf,phase,550.0,npts)

np.savetxt('contour_n168_n138.data',
   np.transpose([cr[0],ci[0],cr[1],ci[1]])) #save to file
plt.plot(cr[0],ci[0],cr[1],ci[1])
plt.xlabel('Real Index')
plt.ylabel('Imaginary Index')
plt.show()
```

15.2.4   SUBROUTINE: TWO_CONTOUR_EQUATIONS.PY

```
import numpy as np
from contour import contour

def two_contour_equations(p,nf1,nf2,ns,na):

    q1,q2 = p
    (cr,ci,_,_,_) = contour(ns,nf1,q1,1.0,2)
    ns = cr[-1]+1j*ci[-1]
    (cr,ci,_,_,_) = contour(ns,nf2,q2,1.0,2)
    return (cr[-1]-np.real(na),ci[-1]-np.imag(na))
```

15.2.5   INTERSECTION BETWEEN TWO CONTOURS (FIGURE 5.4A IN CHAPTER 5)

```
#!/usr/bin/python

import numpy as np
import matplotlib.pyplot as plt
from contour import contour
from two_contour_equations import two_contour_equations
from scipy.optimize import fsolve
#from intersection import intersection

ns = 1.48
npts = 1000                    #Number of points in the film


nf1 = 1.85
nf2 = 1.38
```

```
qr1,qr2 = fsolve(two_contour_equations,(0.2,1.5),(nf1,nf2,ns,1.0))

print qr1,qr2

(cr1,ci1,t1,qr1,qi1) = contour(ns,nf1,qr1,550.0,npts)
ns = cr1[-1]+1j*ci1[-1]
(cr2,ci2,t2,qr2,qi2) = contour(ns,nf2,qr2,550.0,npts)

np.savetxt('contour_n185_n138_upper.data',
 np.transpose([cr1,ci1,cr2,ci2,qr1,t1,qr2,t2])) #save to file

plt.plot(cr1,ci1,cr2,ci2)
plt.xlabel('Real Index')
plt.ylabel('Imaginary Index')
plt.show()
```

15.2.6   SUBROUTINE: VVEQUATIONS.PY

```
import numpy as np

def vvequations(p,ns,nf2,lambda0):
    nf1,delta = p

    t1 = lambda0/2/nf1   #Film 1 thickness
    t2 = lambda0/4/nf2   #Film 2 thickness
    k0 = 2.0*np.pi/lambda0

    q1 = k0*delta*nf1*t1
    nra = nf1 * ((ns + nf1) + (ns - nf1)*np.exp(-1j*2*q1)) \
            / ((ns + nf1) - (ns - nf1)*np.exp(-1j*2*q1))

    q2 = k0*(1.0+delta)*nf2*t2
    na = nf2 * ((nra + nf2) + (nra - nf2)*np.exp(-1j*2*q2)) \
            / ((nra + nf2) - (nra - nf2)*np.exp(-1j*2*q2))
    return (np.real(na-1.0),np.imag(na-1.0))
```

15.2.7   ROOTS OF THE DOUBLE-V DESIGN (FIGURE 5.19A)

```
#!/usr/bin/python

import numpy as np
import matplotlib.pyplot as plt
from scipy.optimize import fsolve
from contour import contour
from vvequations import vvequations

ns = 1.48
nf2 = 1.38
wl = 650.0
nf1,delta = fsolve(vvequations,(1.85,0.3),(ns,nf2,wl))
print nf1
print delta

npts = 1000                     #Number of points in the film
```

```
cr = np.zeros(shape=(2,npts))
ci = np.zeros(shape=(2,npts))
t = np.zeros(shape=(2,npts))
qr = np.zeros(shape=(2,npts))
qi = np.zeros(shape=(2,npts))

phase = 2.0 - 2.0*delta
(cr[0],ci[0],t[0],qr[0],qi[0]) = contour(ns,nf1,phase,wl,npts)
ns = cr[0][-1]+1j*ci[0][-1]
phase = 1.0 - delta
(cr[1],ci[1],t[1],qr[1],qi[1]) = contour(ns,nf2,phase,wl,npts)

np.savetxt('contour_nsolve1_n138_vv.data',
 np.transpose([cr[0],ci[0],cr[1],ci[1]])) #save to file
plt.plot(cr[0],ci[0],cr[1],ci[1])
plt.xlabel('Real Index')
plt.ylabel('Imaginary Index')
plt.show()
```

15.2.8   SUBROUTINE: COMPLEX_NS_EQUATIONS.PY

```
import numpy as np
from contour import contour
def complex_ns_equations(p,ns,na):
    nf,q = p
    (cr,ci,_,_,_,)=contour(ns,nf,q,1.0,2)
    return (cr[-1]-np.real(na),ci[-1]-np.imag(na))
```

15.2.9   SOLVING FOR THE ANTIREFLECTION CONDITION WITH AN EXISTING FILM (FIGURE 5.21)

```
#!/usr/bin/python

import numpy as np
from contour import contour
from complex_ns_equations import complex_ns_equations
from scipy.optimize import fsolve
import matplotlib.pyplot as plt

nf = 2.3
t = 75.0
wl = 550.0
ns = 1.48
phase = 2.0*np.pi/wl*nf*t/(np.pi/2.0)
npts = 1000
cr1,ci1,t1,qr1,qi1 = contour(ns,nf,phase,wl,npts)
ns = cr1[-1]+1j*ci1[-1]
print ''Effective substrate index = '',ns


nf, q =  fsolve(complex_ns_equations,(1.5,np.pi/2.0),(ns,1.0))
print ''Film index = :'',nf,''Phase = '',q
cr2,ci2,t2,qr2,qi2 = contour(ns,nf,q,wl,npts)
```

```
print "Thickness = ",t2[-1]

np.savetxt('contour_n230_nsolve.data',
 np.transpose([cr1,ci1,cr2,ci2,qr1,t1,qr2,t2])) #save to file

plt.plot(cr1,ci1,cr2,ci2)
plt.xlabel('Real Index')
plt.ylabel('Imaginary Index')
plt.show()
```

15.2.10 SOLVING FOR THE METAL AND DIELECTRIC THICKNESSES IN A METAL–INSULATOR–METAL (MIM) STRUCTURE (FIGURE 12.18)

```
#!/usr/bin/python

import numpy as np
from scipy.optimize import fsolve
from contour import contour
from two_contour_equations import two_contour_equations
import matplotlib.pyplot as plt


wl = 550.0 #Wavelength
ns = 1.0
nf1 = 0.125-1j*3.35
t1 = 15.0
qr1 = 2.0*np.pi/wl*np.real(nf1)*t1/(np.pi/2.0)
nf2 = 1.5
nf3 = nf1
npts = 1000

(cr1,ci1,t1,qr1,qi1) = contour(ns,nf1,qr1,wl,npts)
ns = cr1[-1]+1j*ci1[-1]

q2 = 1.0
qr3 = 0.01
q2,qr3 = fsolve(two_contour_equations, (q2,qr3),(nf2,nf3,ns,1.0))
print q2,qr3

npts = 1000
(cr2,ci2,t2,qr2,qi2) = contour(ns,nf2,q2,wl,npts)
ns = cr2[-1]+1j*ci2[-1]

(cr3,ci3,t3,qr3,qi3) = contour(ns,nf3,qr3,wl,npts)


np.savetxt('contour_ag15ag_highT.data',
 np.transpose([cr1,ci1,cr2,ci2,cr3,ci3,
   qr1,qi1,t1,qr2,qi2,t2,qr3,qi3,t3])) #save to file
plt.plot(cr1,ci1,cr2,ci2,cr3,ci3)
plt.xlabel('Real Index')
plt.ylabel('Imaginary Index')
plt.show()
```

Index

A
Absentee, 43, 74
Absorption, 12, 23, 52, 57, 128
Admittance, 11
Ag, 33–34
Al, 35
ALD, 217
Al_2O_3, 26, 66
Amorphous, 32, 198
Anode, 207
Argon, 206
Au, 35

B
Bandgap energy, 32
Blue-shift, 61
Broad band antireflection (BBAR), 76

C
CaF_2, 28
Cathode, 206
CdS, 33
Chalcogenide, 198
Chemical vapor deposition (CVD), 215
Coefficients of thermal expansion (CTEs), 222
Coherence length, 52, 138
Cold mirror, 111
Complex effective index contour, 42
Compressive stress, 222
Cr, 35
Cu, 35

D
Directional power flux, 11, 51
Dispersive, 56, 117
Distributed feedback (DFB), 106
Double-V, 76

E
Effective reflectance index, 41
Electron beam, 214
Equivalent index, 93

F
Fabry–Perot, 112, 125
Face-centered-cubic (FCC), 198
Fresnel reflection, 40, 49

G
Gaussian gradient, 83
Gauss's law, 7
Ge, 33, 87
Geometric mean, 44
Gradient-index, 82
GST, 198

H
Half-wave, 43
Heaviside function, 13
Herpin, 93, 98
Hot mirror, 111

I
Incoherent, 52, 56
Insulator-to-metal transition (IMT), 189
Ion-assisted, 215
ITO, 28, 66

L
LaF_3, 69
Linear gradient, 83
Long pass, 111
Long-wave infrared (MWIR), 23
Loss tangent, 11
Low-E, 184
LPCVD, 216

237

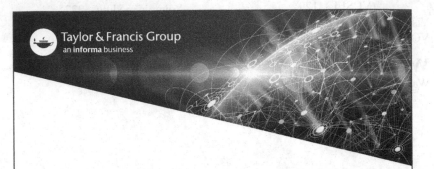

Printed in the United States
by Baker & Taylor Publisher Services